Protocols for Nucleic Acid Analysis by Nonradioactive Probes

Methods in Molecular Biology

John M. Walker, SERIES EDITOR

Methods in Molecular Biology • 28

Protocols for Nucleic Acid Analysis by Nonradioactive Probes

Edited by

Peter G. Isaac

Nickerson BIOCEM Ltd., Cambridge, UK

Humana Press ✳ Totowa, New Jersey

© 1994 Humana Press Inc.
999 Riverview Drive, Suite 208
Totowa, New Jersey 07512

Photocopy Authorization Policy:
Authorization to photocopy items for internal or personal use, or the internal or personal use of specific clients, is granted by The Humana Press Inc., **provided** that the base fee of US $3.00 per copy, plus US $00.20 per page is paid directly to the Copyright Clearance Center at 27 Congress Street, Salem, MA 01970. For those organizations that have been granted a photocopy license from the CCC, a separate system of payment has been arranged and is acceptable to The Humana Press Inc. The fee code for users of the Transactional Reporting Service is: [0-89603-254-X/93 $3.00 + $00.20].

Printed in the United States of America. 9 8 7 6 5 4 3 2

Library of Congress Cataloging in Publication Data

Main entry under title:

Methods in molecular biology.

Protocols for nucleic acid analysis by nonradioactive probes / edited
by Peter G. Isaac.
        p. cm. — (Methods in molecular biology ; 28)
    Includes index.
    ISBN 0-89603-254-X
    1. Nucleic acid probes. 2. Nucleic acids—Analysis. I. Isaac,
Peter G. II. Title: Nonradioactive probes. III. Series: Methods in
molecular biology (Totowa, NJ) ; 28.
QP620.P765 1993
574.19'249—dc20   93-29390
                                                                   CIP

Preface

In assembling this book, *Protocols for Nucleic Acid Analysis by Nonradioactive Probes*, I have endeavored to select protocols that have wide applicability. My aim in doing so is to allow nucleic acid analysis to move from the confines of specialized research laboratories and into general purpose and teaching laboratories. In particular, disciplines such as population biology and clinical diagnostics (where the application of the technology is important, not the technology itself) will find the newer techniques easier to perform with the quantity of samples that are normally required in these fields. Die-hard molecular biologists with hot fingers will find, by trying some of the protocols in this book, that nonradioactive protocols are normally faster than their radioactive brethren, and give more reliable results.

An expanded description of the subject areas covered by this book is given in Chapter 1. The technology herein has very broad application; for instance, in conventional molecular biology research, population biology, plant and animal breeding, genetic mapping, paternity testing, forensics, prenatal diagnosis, and clinical and food microbiology. I have tried to include comprehensive protocols for both basic and more complex analyses. There are chapters dealing with the fundamentals of nucleic acid extraction from plants and animals, and with the procedures necessary to immobilize nucleic acids on solid supports. A range of labeling procedures is included, followed by a selection of hybridization procedures. These are followed by articles on hybridization to chromosomes and to gene transcripts *in situ*. The final chapters of this book give details of novel, high-throughput, nucleic acid screening systems.

I would encourage the reader to browse through *Protocols for Nucleic Acid Analysis by Nonradioactive Probes*, particularly the introduction to each chapter, since information relevant to one probing system will probably be found useful elsewhere. Also, to avoid unnecessary duplication throughout the text, extensive use is made of cross-references. For instance, the synthesis of RNA probes is described in Chapter 12 specifically for minisatellite probes. However, the same technique is used to produce the probes for *in situ* analysis in Chapters 29 and 30. There is a range of nonradioactive systems available, and indications are given in each chapter of the suitability of a protocol for a particular application.

It would have been impossible to produce this book without the constant sage advice from the *Methods in Molecular Biology* series editor, and so I would like to taker this opportunity to express my thanks to Prof. John Walker at University of Hertfordshire. In addition, I would like to thank my colleagues at Nickerson BIOCEM Ltd., Justin Stacey, Liz Davies, and Chris Clee, for coping with my frequent periods of gritted teeth during the production of this book. My colleague Iain Cubitt is also owed an especial debt of thanks for the encouragement he has given me in this project. Finally, I thank the contributors, whose labor made this book possible.

Peter G. Isaac

Contents

vii

Contents

Contributors

ROSA BEDDINGTON • *Centre for Genome Research, University of Edinburgh, Scotland, UK*

PATRICIA M. E. CHADWICK • *Life Sciences Research and Development, Amersham International plc, Amersham, UK*

FARID F. CHEHAB • *Department of Laboratory Medicine, University of California, San Francisco, CA*

JEAN COMPTON • *Cangene Corporation, Mississauga, Ontario, Canada*

MARTIN W. CUNNINGHAM • *Research and Development Division, Amersham International plc, Amersham, UK*

ELIZABETH DAVIES • *Nickerson BIOCEM Ltd., Cambridge, UK*

JOSEPH P. DEL TUFO • *Du Pont Agricultural Products, E. I. du Pont de Nemours and Company, Wilmington, DE*

IAN DURRANT • *Life Sciences Research and Development, Amersham International plc, Amersham, UK*

IAN GARNER • *Pharmaceutical Proteins Ltd., Edinburgh, Scotland, UK*

PAUL A. GRANATO • *Crouse Irving Memorial Hospital, Syracuse, NY*

E. PATRICK GROODY • *Vernon Hills, IL*

BRONWEN M. HARVEY • *Research and Development Division, Amersham International plc, Amersham, UK*

TIM HELENTJARIS • *Department of Plant Sciences, University of Arizona, Tucson, AZ*

J. S. (PAT) HESLOP-HARRISON • *Karyobiology Group, Department of Cell Biology, John Innes Centre, Norwich, UK*

RACHEL HODGE • *Department of Botany, University of Leicester, UK*

PETER G. ISAAC • *Nickerson BIOCEM Ltd., Cambridge, UK*

ANGELA KARP • *Department of Agricultural Sciences, University of Bristol, AFRC Institute of Arable Crops Research, Long Ashton Research Station, Bristol, UK*

ERNEST S. KAWASAKI • *Procept Inc., Cambridge, MA*

xi

JON KRATOCHVIL • *Abbott Laboratories, Diagnostics Division, North Chicago, IL*

THOMAS G. LAFFLER • *Abbott Laboratories, Diagnostics Division, North Chicago, IL*

ANDREW R. LEITCH • *Karyobiology Group, Department of Cell Biology, John Innes Centre, Norwich, UK*

ILIA J. LEITCH • *Karyobiology Group, Department of Cell Biology, John Innes Centre, Norwich, UK*

LARRY MALEK • *Cangene Corporation, Mississauga, Ontario, Canada*

TOM MCCREERY • *Department of Plant Sciences, University of Arizona, Tucson, AZ*

JULIA M. POLAK • *Histochemistry Department, Royal Postgraduate Medical School, Hammersmith Hospital, London, UK*

BARRY ROSEN • *Centre for Genome Research, University of Edinburgh, Scotland, UK*

TRUDE SCHWARZACHER • *Karyobiology Group, Department of Cell Biology, John Innes Centre, Norwich, UK*

ESTHER N. SIGNER • *Department of Genetics, University of Leicester, UK*

ROY SOOKNANAN • *Cangene Corporation, Mississauga, Ontario, Canada*

JUSTIN STACEY • *Nickerson BIOCEM Ltd., Cambridge, UK*

TIMOTHY STONE • *Life Sciences Research and Development, Amersham International plc, Amersham, UK*

GIORGIO TERENGHI • *Histochemistry Department, Royal Postgraduate Medical School, Hammersmith Hospital, London, UK*

SCOTT V. TINGEY • *Du Pont Agricultural Products, E. I. du Pont de Nemours and Company, Wilmington, DE*

CLAIRE B. WHEELER • *Research and Development Division, Amersham International plc, Amersham, UK*

CHAPTER 1

About Nonradioactive
Nucleic Acid Detection

Peter G. Isaac

1. The Need for Nonradioactive Systems

It is the aim of this book to provide working, reliable, protocols for nucleic acid analysis that avoid the use of radioisotopes. It is hoped that in doing so, the widespread use of this technology will be encouraged in areas where the older procedures, based on radioisotopes, are not practicable.

Science moves by paradigm shifts in technology. The publications in 1975 by Southern *(1)* and Grunstein and Hogness *(2)* started a scientific revolution that largely continues unabated. These papers describe the core technologies that enable specific sequences of DNA to be identified within a mixture of other DNA fragments *(1)*, or within a plasmid harbored by a transgenic bacterium *(2)*. This technology enabled the genomes of higher organisms to be investigated in ways that were hitherto impossible. The technique of probing immobilized nucleic acid mixtures with a characterized DNA sequence enables the establishment of the number of copies of a particular gene sequence present in an organism, detailed genetic investigation via restriction fragment length polymorphism (RFLP) analysis, and the means to test if potentially transgenic organisms have acquired a new gene. The use of probes to examine the organization of chromosomes,

From: *Methods in Molecular Biology, Vol. 28:*
Protocols for Nucleic Acid Analysis by Nonradioactive Probes
Edited by: P. G. Isaac Copyright ©1994 Humana Press Inc., Totowa, NJ

and the patterns of expression of genes in tissue sections, developed in parallel with these studies.

Since the mid 1970s, the technology itself has changed remarkably little in principle, but one single overriding problem has prevented nucleic acid identification technology from moving from a highly specialized research setting to areas where it still has great promise. The use of radioisotopes has meant that the everyday technology of the molecular biologist remains the province of those who have access to laboratories that are registered and maintained for the use of radioisotopes. The concern over safe working practices with radioisotopes is only one feature that has hampered the widespread adoption of nucleic acid identification technology in applied settings. The hidden costs of using radioisotopes (from a United Kingdom perspective) are discussed in the next few paragraphs.

The limits on storage and disposal of radioisotopes have always placed an artificial ceiling on the number of assays that can be performed routinely by any laboratory. A plant breeding program generates hundreds of thousands of different genotypes of plants per year. It is clearly impractical to screen all of these using many probes (for instance, as RFLP markers), but even to screen a fraction of the plants implies that the technology must be robust enough for routine large scale screening. The limitations on isotope usage mean that RFLPs are applied to only a tiny percentage of the plants.

Laboratories that are used for radioisotope work in the United Kingdom must either be registered as Supervised or Controlled areas. Both imply restricted access, separate storage facilities for radioisotopes, and careful management of the use to which the laboratory is put. To maintain a separate laboratory away from the main working laboratory is usually an expensive option as it implies extra rent payable on the occupied space, and usually the duplication of equipment already present in a main laboratory. Registration to hold and dispose of isotopes requires both an initial payment to Her Majesty's Inspectorate of Pollution, plus a yearly retainer.

In order to work with radioisotopes, staff must be given training in the safe use of radioisotopes. In addition, in the United Kingdom, additional training is required for those responsible for the day-to-day maintenance and inspection of radioisotope facilities (the Radiation Protection Supervisor). Any establishment involved in work with ionizing radiation must also appoint a Radiation Protection Adviser (usually an independent consultant). All this costs money, and more importantly, time.

In our laboratory, we perform RFLP analysis on many species of plants. Plants tend to have large genomes, and in order to get a reasonable signal from a radioisotopically (^{32}P) labeled probe hybridized to a single-copy-per-genome sequence, it is usually necessary to expose an autoradiograph for 14 d. Using the digoxigenin and AMPPD system described in this book (*see* Chapter 17) we now obtain signals within a few hours. In general, the result has a clearer background and sharper bands than can be obtained with ^{32}P. As a result, through adopting a nonradioactive probing technology, fewer hybridization filters have to be made than previously, and we do not have the additional costs of buying intensifying screens for every blot that is exposing at any given time.

As a plant breeding company, we have an interest in disseminating an awareness of nucleic acid identification technology to plant breeders throughout the world. The use of radioisotopes is a barrier to the adoption of this technology, mainly because of the "bad press" of radiation. The result is a rejection of the technology that has enormous potential to plant breeders in their craft of plant genetics.

The widespread use of nonradioactive methods for measuring genetic diversity has great potential, especially in the Third World. Such measurements, based upon the analysis of nucleic acid variation, allow the evaluation of the effectiveness of conservation strategies and could also lead to the development of food crops containing optimal genotypes for their survival in harsh environments. But, with heroic exceptions, scientists in the Third World do not have access to short lived radioisotopes, nor the facilities with which to deal with them safely.

Since the early 1980s, it has been practical to perform nucleic acid hybridizations using nonradioactive systems *(3)*, instead of radioisotopes, which were used previously. Recent advances in substrate chemistry and probe labeling have dramatically increased the sensitivity of the nonradioactive methods, and these form the basis of this book.

In parallel with the adoption of nonradioactive probing, there has been a general realization that the gel-filter-probe-expose cycle, although adequate for a research laboratory dealing with a few tens of samples, is not well adapted to handling thousands of samples. Consequently, some newer technologies have been developed so that once a nucleic acid sequence is known its presence can be detected in rapid and nonradioactive ways. Some of these procedures are described in the final chapters of this book.

2. About This Book

The protocols given in this book are divided into seven sections. The following paragraphs give an overview of the contents of these chapters.

2.1. DNA Preparation and Blotting

Chapters 2 and 3 describe the procedures necessary to isolate plant and animal DNA of a suitable grade for restriction analysis, blotting, and probing with nonradioactive probes. Chapter 4 describes the procedures necessary to convert this high-mol-wt DNA into restriction fragments and how to blot it, using a vacuum system, to prepare a filter ready for probing. The protocols described in these chapters are those that are necessary to produce membrane filters for RFLP analysis, checking for the insertion of transformed genes into host chromosomes, checking for gene copy number and so on.

2.2. RNA Preparation and Blotting

Chapters 5 and 6 describe the preparation of plant and animal RNA. Chapter 6 also gives details on the further purification of the mRNA (i.e., polyadenylated) fraction from the total RNA. Chapter 7 describes the procedures necessary to separate this RNA by gel electrophoresis, and to transfer the separated fragments onto a nylon membrane. Chapter 8 describes a protocol for producing RNA dot blots on a nylon membrane. The protocols described in these chapters are those necessary to produce membrane filters for determining levels of expression of endogenous, or newly introduced, genes and the sizes of gene transcripts.

2.3. Probe Preparation

Chapter 9 describes the basic procedure necessary to make a plasmid, as this is a common starting point for the preparation of a probe. Chapters 10–16 describe how to make nonradioactive probes labeled with digoxigenin, biotin, horseradish peroxidase, and fluorescein. The protocols include using the polymerase chain reaction (PCR, Chapter 10), random hexanucleotide priming (Chapters 11 and 15), transcript labeling (Chapter 12), nick translation (Chapter 13), direct enzyme labeling of DNA using glutaraldehyde (Chapter 14), and tailing of oligonucleotides (Chapter 16).

2.4. Use of Probes on Blots

Chapters 17–22 describe the hybridization of the nonradioactive probes to the DNA and RNA immobilized on blots, together with the detection systems necessary to reveal where the probe has hybridized. Chapters 17–19 deal with digoxigenin probes, with Chapters 17 and 19 describing chemiluminescent detection on DNA and RNA blots respectively, and Chapter 18 describing a colorimetric detection system. Chapter 20 deals with enhanced chemiluminescent detection of enzymically labeled probes, whereas Chapters 21 and 22 describe enhanced chemiluminescent detection of large (Chapter 21) and small (oligonucleotide, Chapter 22) probes labeled with fluorescein.

2.5. Hybridization to Chromosomes **In Situ**

Chapters 23–27 describe the preparation of plant chromosome spreads and their use for hybridization of probes to chromosomes *in situ*. Chapters 23 and 24 describe different methods of preparing plant chromosome spreads. The use and detection of the probes is described for probes labeled with biotin (Chapter 25), fluorescent probes (Chapter 26) and digoxigenin probes (Chapter 27). The protocols described in these chapters are used in situations where multiple copy sequences can be used to identify chromosome arms (i.e., as a cytogenetic paintbrush) and in situations where it is essential to identify the presence or absence of alien chromosomes or chromosome arms in a host cell.

2.6. Hybridization to RNA **In Situ**

Chapters 28 and 29 describe the use of nonradioactive probes to detect gene transcripts in thin sections *in situ*. Chapter 30 describes the use of nonradioactive probes on whole embryo mounts. The probes used in this section are digoxigenin-labeled RNA probes, and the detection is colorimetric, revealing the cell and tissue types that are expressing particular genes.

2.7. New Methods: The Window on the Future

The remaining Chapters, 31–36, cover newer kinds of technologies, where the problem under investigation is not confined to a research setting and a small number of samples. These systems have as their goal easy and high throughput screening for the presence of particular nucleic acid sequences.

Chapters 31 and 32 deal with similar concepts, in that probes are hybridized to a nucleic acid extract, derived by bacterial lysis. The probes have a label that can be detected in a nonradioactive assay, and the probe plus target complex can be specifically captured by magnetic beads (Chapter 31) or a plastic dipstick (Chapter 32). A positive signal means that a sequence, and hence a particular bacterium, is present in the tested sample. These protocols are useful in clinical and food microbiological contexts, where diagnosis of bacterial presence must be made quickly, and can be used to formulate a decision whether or not other confirmatory tests should be carried out.

In the reverse dot blot method (Chapter 33), a locus from an individual is amplified by a polymerase chain reaction using biotinylated primers. This product is used to probe allele-specific sequences immobilized on a nylon membrane. A positive signal on some of the dots implies the presence of particular alleles of the tested locus. This technology can been used in clinical applications (for testing for the presence of particular histocompatibility alleles before organ grafting), forensic analysis (determining the histocompatibility alleles present in suspect blood samples), and other aspects of human genetics (as in the example given in Chapter 33, demonstrating the analysis of the cystic fibrosis locus). With these applications, diagnosis has to be rapid, accurate, and sensitive to single, or a few, nucleotide differences between alleles of the same locus.

Chapters 34–36 deal with different ways of using amplification methods to reveal polymorphism within nucleic acids. The RAPD technique (Chapter 34) uses short primers (normally 10 bases) in a PCR reaction on two or more DNA samples—several bands are produced and differences in the banding patterns infers the presence of different sequences in the two tested DNAs. These polymorphic bands behave as dominant genetic markers, and the technique is useful if a large number of markers must be found to distinguish a few (usually two) genotypes.

The ligase chain reaction (LCR, Chapter 35) uses four oligonucleotides in a ligation reaction with a thermostable ligase and a DNA template. The sequences of these oligonucleotides are chosen such that ligation only occurs where the oligonucleotides match the DNA template perfectly. Where there is no match, ligation does not occur—as a result LCR can be used to detect the presence of particular alleles. Chapter 35 also includes information on how to detect the products of the LCR reaction without resorting to gel electrophoresis.

Chapter 36 describes a protocol for an isothermal chain reaction that makes use of the presence of DNA primers, T7 promoters, T7 polymerase, RNaseH, and AMV reverse transcriptase to produce large quantities of nucleic acid when the primers used match an RNA (or DNA) template. This system has the advantage that no thermal cycling block is required, and hence the number of samples that can be processed daily is not limited by the availability of specialized equipment.

References

1. Southern, E. M. (1975) Detection of specific sequences among DNA fragments separated by gel electrophoresis. *J. Mol. Biol.* **98,** 503–517.
2. Grunstein, M. and Hogness, D. S. (1975) Colony hybridization: A method for the isolation of cloned DNAs that contain a specific gene. *Proc. Natl. Acad. Sci. USA* **72,** 3961–3965.
3. Leary, J. J., Brigati, D. J., and Ward, D. C. (1983) Rapid and sensitive colorimetric method for visualizing biotin-labeled DNA probes hybridized to DNA or RNA immobilized on nitrocellulose: Bio-blots. *Proc. Natl. Acad. Sci. USA* **80,** 4045–4049.

CHAPTER 2

Isolation of DNA from Plants

Justin Stacey and Peter G. Isaac

1. Introduction

This chapter describes a DNA extraction method that can be used both on freeze-dried leaves and on fresh leaves, and is based on the method of Saghai-Maroof *(1)*, modified by David Hoisington and Jack Gardiner at the University of Missouri at Columbia (personal communication). The scale of the extraction is dependent on the amount of starting material; 300–400 mg freeze-dried material requires 9 mL extraction buffer and should yield 250 μg to 1 mg DNA. Scaling the procedure down 10-fold results in a miniprep method that is very suitable for extracting small quantities (25–100 μg) of DNA from young, fresh leaves.

The DNA is not free from contaminants such as carbohydrates, but is of a suitable grade for enzyme digestion, Southern blotting *(2,* and Chapter 4), and analysis by polymerase chain reaction (PCR), such as the RAPD technique *(3,* and Chapter 34). We have used this procedure on wheat, barley, maize, oilseed rape, vegetable Brassicas, peas, and onions. A benefit of the method is that it requires little "hands-on" time by the operator, and can therefore be used to process large numbers of samples on a daily basis.

One of the problems associated with making DNA of sufficient quality for Southern blotting and PCR analysis is that DNA can be sheared if it is manipulated too violently. It is therefore important

From: *Methods in Molecular Biology, Vol. 28:*
Protocols for Nucleic Acid Analysis by Nonradioactive Probes
Edited by: P. G. Isaac Copyright ©1994 Humana Press Inc., Totowa, NJ

that any DNA extraction method uses the minimum number of mixing steps. Solutions tend to be mixed with the DNA by gentle inversions of the tube. For the same reason, it is important to avoid transferring the DNA solution too many times between fresh tubes. Ideally the solution should be transferred by careful pouring, but, if that is not possible, the DNA should be transferred using an inverted sterile pipet (so that the DNA solution is drawn into what is normally the mouthpiece) or a pipet tip widened by cutting off the fine point.

The procedure described in Section 3. is for 300–400 mg of freeze-dried leaf material. A detergent (cetyltriethylammonium bromide, CTAB) is used to break open plant cells and solubilize the contents. Chlorophyll and some denatured proteins are removed from the green plant tissue in an organic chloroform/octanol step, and the organic phase is separated by a brief centrifugation. At this point the extract contains RNA and DNA, and the former is removed by incubating with RNase A. The DNA is precipitated and washed in organic solvents before redissolving in aqueous solution. The concentration of the DNA is then estimated by spectrophotometry and agarose gel electrophoresis.

2. Materials

1. $1M$ Tris-HCl, pH 8.0: Filter and autoclave.
2. $5.0M$ NaCl: Filter and autoclave.
3. $0.5M$ EDTA (disodium ethylenediaminetetraacetic acid): Weigh out an appropriate amount of EDTA and add to stirring distilled water (about 75% of the final volume). Add sodium hydroxide pellets slowly until the solution begins to clear. Monitor the pH and add NaOH until the EDTA has dissolved and the pH reaches between 7 and 8. Make up to the final volume, filter through a paper filter, and autoclave. Store the solution in a refrigerator.
4. CTAB extraction buffer: For 100 mL (enough for 10 isolations) mix 73 mL deionized water, 10 mL $1M$ Tris-HCl, pH 7.5 , 14 mL $5M$ NaCl, and 2 mL $0.5M$ EDTA, pH 8.0. This solution should be filtered and autoclaved. The solution can be stored on the bench at room temperature. Immediately prior to use add 1 mL of β-mercaptoethanol and 1 g CTAB (*see* Note 1). Preheat the solution to 65°C.
5. Chloroform:octanol (24:1). Store in dark at room temperature. Make up and dispense this solution in a fume cupboard.

6. Preboiled RNaseA (10 mg/mL): Dissolve RNaseA in water, place the tube in a boiling water bath for 10 min, and allow to cool on a bench. Store at –20°C.
7. Isopropanol.
8. 3*M* Sodium acetate, pH 6.0: Adjust the pH with acetic acid before making to the final volume. Filter and autoclave this solution and store at room temperature.
9. 76% Ethanol, 0.2*M* sodium acetate: For 100 mL (10 isolations) mix 76 mL absolute ethanol, 6.7 mL 3*M* sodium acetate pH 6.0, and 17.3 mL of autoclaved deionized water. Store at 4°C until ready for use.
10. 70% Ethanol. Store at –20°C.
11. 1*M* Tris-HCl, pH 8.0.
12. TE buffer: For 100 mL mix 98.8 mL deionized water, 1 mL 1*M* Tris-HCl, pH 8.0, 0.2 mL 0.5*M* EDTA, pH 8.0. Filter and autoclave solution.
13. Freeze-drier bags: We use bags made from heat-sealable tea bag paper from Crompton Ltd. A package of approx 10 × 30 cm is made by heat sealing three edges of a doubled over 10 × 60 cm length, thus making a pocket with an open end. The bags are held closed with paper clips.
14. Miracloth (Calbiochem, San Diego, CA).
15. Glass hooks are made from Pasteur pipets by placing about 5–10 mm of the fine end of the pipet horizontally in a Bunsen flame, so that the end becomes sealed. The end of the pipet will slowly droop under gravity. Remove the pipet from the flame and hold it pointing vertically (bent end upward). The molten glass will form a hook.
16. A sample mill for grinding material, e.g., a Tecator cyclotec 1093, fitted with the finest sample mesh. Alternatively a pestle and mortar plus quartz sand can be used.

3. Method

1. Place freshly harvested plant leaf samples in labeled freeze-drier bags. Close the bags with paper clips then place the samples in a –80°C freezer (*see* Note 2), and leave until frozen (longer than 6 h). Transfer the bags to a freeze-drier, evacuate the chamber, and freeze-dry overnight or until the samples are dry (*see* Notes 3 and 4). At this point the leaves should be uniformly brittle.
2. After removing pieces of stem and leaf midribs (*see* Note 5), mill the samples using either a sample mill or pestle and mortar with grinding sand. Use a fresh pestle and mortar for each sample, or, if using a mill, thoroughly clean the apparatus (using a brush and vacuum cleaner) before processing the next sample (*see* Note 6).

3. To 300–400 mg lyophilized ground tissue in a sterile, disposable 16-mL polypropylene centrifuge tube, add 9 mL prewarmed CTAB extraction buffer (*see* Note 7). Mix gently by inversion.
4. Incubate the samples for 60–90 min, with occasional inversion at 65°C.
5. Allow the samples to cool by standing the tubes in a trough of water at room temperature for 5 min.
6. Add 5 mL chloroform:octanol (24:1). Rock the tubes gently (or rotate them on a tube roller) to mix for 5 min.
7. Spin the samples in a bench-top centrifuge for 2 min at 850g and room temperature.
8. Pour off the top (aqueous) layer into a fresh 16-mL tube (*see* Note 8) and add 50 µL of preboiled RNaseA (10 mg/mL). Mix the samples gently by inversion and incubate for 30 min at room temperature.
9. Add 6 mL of isopropanol to each tube. Mix the samples gently by inversion until a white fluffy DNA precipitate appears (it should appear within about 1 min, *see* Note 9).
10. After 2–3 min remove the precipitated DNA with a glass hook (*see* Note 9) and transfer to a fresh 16-mL tube containing 8 mL of cold 76% ethanol, 0.2M sodium acetate. Leave the DNA on the hook in the tube for 20 min.
11. Transfer the DNA to a fresh 16-mL tube containing 8 mL of cold 70% ethanol for a few seconds then transfer the DNA to a fresh 16-mL tube containing 1 mL TE.
12. Rock gently to disperse DNA. Once the DNA has detached from the glass hook the hook can be removed from the tube. Leave the samples at 4°C overnight to allow the DNA to dissolve (*see* Note 10).
13. Calculate the DNA concentration by measuring the absorbance at 260 nm of a small aliquot of the sample in quartz cuvets in a UV spectro-photometer. A solution of 50 µg/mL has an optical density of 1 in a 1-cm cuvet.
14. The integrity of the DNA can be visualized by running a 10-µL aliquot of the sample on a low percentage (<0.7%) agarose gel stained with ethidium bromide (*see* Chapter 4). The DNA should appear as a high-mol-wt band running with, or slower than, a 20-kbp size marker (such as the largest *Hin*DIII fragment of bacteriophage lambda). (*See* Notes 11–14.)

4. Notes

1. Some grades of CTAB do not appear to work as well as others. We use a preparation called mixed alkyltrimethylammonium bromide, available from Sigma. Poor yields of DNA have resulted when another preparation was used.

2. Samples can also be placed in layers of dry ice (solid CO_2) pellets. This is a particularly useful method when samples are being collected from the field.

3. The chamber temperature on the freeze drier can be left at ambient (in fact, cooling the sample chamber increases the drying time).

4. Freeze-drying is not the only method that can be used to dry plant material, although the integrity of the DNA may be compromised. Other methods include air-drying at 65°C in an incubator with forced air circulation or drying under vacuum. In each case the samples should be dried to constant weight. In the case of air-drying this is usually overnight.

5. Some species, for instance maize and lettuce, have very pronounced leaf midribs, and these should be removed from the material before grinding. Removal of the midrib is not important if it is small or non-existent, e.g., in very young leaves or in small grain cereals. The reason why the midrib must be removed in some cases is that it is a major source of carbohydrate contamination.

6. Other methods of disrupting the plant tissue are available. For some species, e.g., rice and other cereals, a domestic coffee mill *(4)* can also be used, provided that a reasonable amount of dried leaf is to be ground. Alternatively, the sample can be placed in a 50-mL polypropylene centrifuge tube with glass beads. The sample can then be vigorously agitated using a paint mixer for 0.5–3 min *(4)*.

7. DNA can be extracted from fresh plant tissue by grinding a leaf or leaf disc in a small amount of extraction buffer. If you are extracting from leaf discs, a scaled down miniprep isolation procedure can be done as follows. Grind about 1 cm^2 of fresh leaf in 0.5 mL of extraction buffer using a glass rod in a small polystyrene chemical weighing boat. Pour this into a 1.5-mL centrifuge tube, wash the weighing boat out with a further 0.5 mL of extraction buffer, and pool with the first extraction. Continue from step 4, dividing the reagent volumes by 10.

8. Occasionally, leaf debris is not packed tightly enough into a solid plug separating the organic and aqueous layers. When this occurs two things can happen. First, the whole contents of the tube can slop into the fresh tube. If this occurs, the sample must be respun, and the aqueous phase drawn slowly up through a 1-mL Gilson-type pipet tip that has had the fine end trimmed off to make it into a wider bore; this aqueous phase should be transferred to a fresh tube with fresh RNaseA. Second, a few small pieces of leaf can contaminate the new tube. If this occurs, pour the solution through Miracloth into a fresh tube.

9. After adding isopropanol the DNA may not form a clot. Instead it forms several smaller fragments that are very difficult to remove using a glass hook (this is usually the case with the miniprep scale). In this instance centrifuge at 850g (or microcentrifuge for the miniprep) for 5 min to pellet the DNA. Wash the pellet with the solutions described in steps 10 and 11, repelleting after each wash. Finally resuspend the DNA in TE.

10. If the DNA solution appears turbid after standing overnight at 4°C, try heating the sample to 65°C for 10 min, inverting the tube every 3 min. Insoluble material that remains after this treatment can be removed by centrifugation at 850g for 5 min, and the cleared supernatant can be removed to a fresh tube. The pellet can be discarded (it is not DNA).

11. When checking the integrity of the DNA on an agarose gel an estimate of the relative concentrations of the samples can be made by viewing the intensity of the DNA bands, particularly if DNA mol-wt markers of known concentration are run on the same gel. This, combined with the measurement of the optical density of the DNA in a spectrophotometer, gives a more reliable estimate of the concentration. For RFLP and PCR analysis it is important that each sample is at the same concentration.

12. Should the DNA appear degraded (i.e., as a smear running down the gel), an isolation made from fresh plant tissue may yield intact DNA. When harvesting plant material for freeze-drying, ensure that the tissue is immediately frozen as this reduces DNA degradation. In addition, making fresh solutions, particularly RNaseA, may cure the problem. Finally, DNA is a large molecule that can be broken by shear forces if treated with too much violence. Therefore care should be taken to mix samples gently, never vortex the DNA.

13. Occasionally despite having an optical density at 260 nm, there appears to be no DNA on the gel, but instead the sample well glows brightly. The presence of DNA-protein aggregates often prevents the DNA from moving into the gel (*see* Note 14). Digesting the sample with protein-ase will remove the protein, releasing the DNA. The sample should be subsequently reextracted by performing steps 3–14 or can be phenol extracted (*see* Chapter 3).

14. The DNA should not be allowed to dry at any stage during the preparation as this hinders resuspension and solubilization in TE. This may be because the DNA and residual denatured proteins form an insoluble mass.

References

1. Saghai-Maroof, M. A., Soliman, K. M., Jorgensen, R. A., and Allard, R. W. (1984) Ribosomal DNA spacer-length polymorphisms in barley: Mendelian inheritance, chromosomal location and population dynamics. *Proc. Natl. Acad. Sci. USA* **81,** 8014–8018.

2. Southern, E. M.(1975) Detection of specific sequences among DNA fragments separated by gel electrophoresis. *J. Mol. Biol.* **98,** 503–517.

3. Williams, J. G. K., Kubelik, A. R., Livak, K. J., Rafalski, J. A., and Tingey, S. V. (1990) DNA polymorphisms amplified by arbitrary primers are useful as genetic markers. *Nucleic Acids Res.* **18,** 6531–6535.

4. Tai, T. H. and Tanksley, S. D. (1990) A rapid and inexpensive method for isolation of total DNA from dehydrated plant tissue. *Plant Mol. Biol. Reporter* **8,** 297–203.

CHAPTER 3

Isolation of High-Molecular-Weight DNA from Animal Cells

Ian Garner

1. Introduction

Mammalian chromosomes are of the order of 12–60 times the size of that of *Escherichia coli* (4×10^3 kilobase pairs [kbp]) *(1)*. The choice of method used when purifying DNA from mammalian cells may be dictated by the use to which the product will be put as it will influence the average size of the material purified. For example, methods incorporating many aggressive manipulations will tend to shear the DNA into molecules of relatively low-mol-wt (< 50 kbp). This may be suitable for polymerase chain reaction (PCR) *(2,3)* analysis and in some cases Southern blotting *(4)* but will be unsuitable for other more demanding purposes, e.g., genomic library constructions. When performed with care, methods involving minimal manipulations will yield DNA in excess of 200 kbp, suitable for most purposes. In general, it is prudent to utilize such methods for all preparations of DNA from mammalian cells. The first three methods described below are derived from that of Blin and Stafford *(5)* and should yield high-mol-wt (HMW) DNA from solid tissues, blood, or cells in culture suitable for most purposes including cloning, PCR/RFLP analysis, and Southern blotting. The final method described is

From: *Methods in Molecular Biology, Vol. 28:*
Protocols for Nucleic Acid Analysis by Nonradioactive Probes
Edited by: P. G. Isaac Copyright ©1994 Humana Press Inc., Totowa, NJ

that of Lahiri and Nurnberger *(6)* and is a rapid approach that eliminates the use of solvents and enzymes, making it easier to process
large numbers of samples. The material produced by this method
should be approx 50 kbp and is suitable for PCR and RFLP analysis.

2. Materials

2.1. Preparation of HMW DNA from Solid Tissues

If possible, all materials should be sterilized prior to use.

1. Liquid nitrogen.
2. Porcelain pestle and mortar prechilled to –20°C.
3. Selection of spatulas.
4. 600-mL Pyrex beakers (or similar wide-based vessel).
5. TE: 10 mM Tris-HCl, pH 8.0, 1 mM EDTA.
6. Phenol saturated with TE *(see* Note 1)
7. Extraction buffer: 0.1M EDTA, 0.2M NaCl, 0.05M Tris-HCl, pH 8.0, 0.5% SDS, 50 μg/mL DNase-free RNase *(see* Note 2).
8. Proteinase K: 20 mg/mL in sterile distilled water.
9. Dialysis tubing (wide bore): Preboiled in 1 mM EDTA and rinsed with sterile distilled water.
10. 3M Sodium acetate adjusted to pH 6.0 with acetic acid.
11. Absolute ethanol.
12. 70% Ethanol.
13. Pasteur pipets with sealed hooked ends.
14. Pasteur pipets attachable to a vacuum line.

2.2. Preparation of HMW DNA from Blood

Materials as for Section 2.1. with the addition of the following:

1. Heparinized Vacutainers.
2. Hanks buffered saline (HBS) (from Sigma, St. Louis, MO).
3. Histopaque (from Sigma).

2.3. Preparation of HMW DNA from Cells in Culture

Materials as for Section 2.1. with the addition of the following:

1. Ca^{2+}/Mg^{2+} free Phosphate buffered saline (PBS) from Gibco (Grand Island, NY) or similar supplier.
2. Rubber policemen or similar cell scrapers.

2.4. Preparation of HMW DNA from Blood Without the Use of Solvents or Enzymes

Materials 5,11,12, and 13 from Section 2.1. with the addition of the following:

1. Low salt buffer: 10 mM Tris-HCl, pH 7.6, 10 mM KCl, 10 mM MgCl$_2$, 2 mM EDTA.
2. High salt buffer:10 mM Tris-HCl, pH 7.6, 10 mM KCl, 10 mM MgCl$_2$, 0.4M NaCl, 2 mM EDTA.
3. Nonidet P-40.
4. 10% SDS.
5. 6M NaCl.

3. Methods

3.1. Preparation of HMW DNA from Solid Tissues

1. Freshly excised tissues should be dropped immediately into liquid nitrogen. Large organs should be cut into smaller, more manageable pieces (<1 cm^3) as this will ease freezing, storage, and subsequent manipulations. Any organ can be taken, but should liver be required, a 24-h starvation period prior to sacrifice will improve DNA quality. Tissues harvested in this way can be stored at –70°C for several years prior to use.
2. Pour liquid nitrogen into the precooled mortar, add the tissue of choice (up to 1 cm^3), and grind to a fine powder with the pestle. It may be necessary to break up the tissue into smaller pieces to facilitate grinding (e.g., by wrapping in tin foil and hitting with a hammer). Add more liquid nitrogen as required to keep the sample cold.
3. Once the sample has been ground, allow the liquid nitrogen to evaporate and use a spatula to transfer the powdered tissue to the surface of 20 mL of extraction buffer in a 600-mL beaker at room temperature. Sprinkle the powder evenly over the surface of the liquid and gently swirl the beaker to submerge the material.
4. Add Proteinase K to 100 µg/mL (100 µL of stock) and gently swirl the beaker to mix the components. Incubate the beaker at 37°C for at least 3 h, preferably overnight, with gentle agitation. This can be achieved using a shaking water bath or by occasional swirling by hand. The solution should be reasonably clear and viscous at the end of the incubation. More Proteinase K may be added to achieve this (*see* Note 3).
5. Add 20 mL of equilibrated phenol (*see* Note 1) and seal the beaker with parafilm. Gently swirl by hand for 10–15 min to mix the two phases.

The larger the surface area available, the easier this will be. Ideally you should generate an emulsion at this stage. It may be necessary to transfer the mixture to a larger container to achieve this.

6. Transfer the mixture to a 50-mL disposable plastic tube and centrifuge at 1500g for 10 min at room temperature to separate the two phases.

7. Remove the lower phenol phase by gentle aspiration through a Pasteur pipet attached to a vacuum line through a side arm flask. The pipet should be lowered into the lower phenol phase with the vacuum line clamped until the thread of viscous DNA has detached from the pipet tip. Slowly unclamp the vacuum line and allow the phenol phase to run into the flask. Once all of the phenol has been removed, the vacuum line is again clamped and the pipet is removed. Alternatively, the aqueous phase can be removed with a wide-bore pipet. However, care must be taken not to disturb the interface and when the DNA is very viscous this is hard to achieve (*see* Note 4).

8. Steps 5–7 should be repeated to accomplish at least three extractions. The aqueous phase should be clear at this point.

9. At this stage two routes to recovery of HMW DNA are available (*see* Note 5).

   a. Dialyze the aqueous phase against 1000 vol of TE. This should be performed for 30 min at room temperature to prevent SDS precipitation in the sample followed by overnight at 4°C. Allow room for expansion in the dialysis bag.

   b. Transfer the aqueous phase to a fresh beaker and add sodium acetate to 0.3M. Mix by gentle swirling. Add 2 vol of absolute ethanol and mix by gentle swirling. The DNA will begin to precipitate almost immediately in a strandy complex. Initially this will be glass-like but it will begin to attain a white appearance as the precipitation proceeds. Hook out the DNA strands using a Pasteur pipet with a sealed U-shaped end before they attain too white an appearance (*see* Note 6). Dip the DNA in 70% ethanol for a few seconds and allow to air dry for a few minutes. Transfer the DNA to 1–3 mL TE. Gently wet the DNA in the liquid and allow it to fall off the pipet tip onto the surface of the liquid. Do not shake violently to achieve this. Leave to dissolve overnight at 4°C. If this proves difficult, incubate the tube overnight at room temperature on a gently rocking table or rotating wheel (*see* Note 4)

10. The absorbance of the DNA at 260 nm and 280 nm should be measured using quartz cuvets. The 260/280 ratio should be >1.8. If this is not the case, repeat steps 4–9 adding additional SDS to 1%. An A_{260} of 1.0 in a 1-cm light path is equivalent to a DNA concentration of 50 µg/mL. Store the DNA at 4°C.

11. An aliquot of the DNA should be analyzed by electrophoresis through a 0.3% agarose gel. Multimers of bacteriophage lambda generated by ligation or commercially available DNAs can serve as mol-wt markers. The prepared DNA is normally at least 100 kbp and preferably exceeds 200 kbp (*see* Notes 7–10).

3.2. Preparation of HMW DNA from Blood

1. Blood should be collected into heparinized vacutainers. Ideally, it should be processed immediately but can be stored overnight at 4°C.
2. Dilute 10 mL of blood with 10 mL of HBS.
3. Layer this over 5 mL Histopaque and centrifuge in a 15-mL disposable plastic centrifuge tube for 15 min at room temperature at 2000*g*.
4. A white band containing peripheral lymphocytes should be visible in each tube. Remove and discard the sample above this and transfer the white band to a fresh 15-mL centrifuge tube.
5. Wash the cells by adding 10 mL of HBS, mix thoroughly, and recover the cells by centrifugation for 10 min at room temperature at 2000*g*.
6. Discard the supernatant and resuspend the cell pellet in 20 mL extraction buffer. Continue as from Section 3.1., step 4.

3.3. Preparation of HMW DNA from Cells in Culture

1. Cells ($\sim 10^8$) should be grown as a monolayer or in suspension as required (*see* Note 9).
2. For monolayers, decant the medium and rinse the cells twice with PBS. Recover the cells by scraping with a rubber policeman and centrifuge at 500*g* for 10 min in a 15-mL plastic disposable centrifuge tube at room temperature. For cell suspensions, transfer to 15-mL plastic disposable centrifuge tubes and recover the cells by centrifugation for 10 min at room temperature at 500*g*. Resuspend the cell pellet(s) by gentle pipeting in 10 mL PBS and recover by centrifugation as above.
3. Resuspend the cell pellet in 20 mL extraction buffer and continue as from Section 3.1., step 4.

3.4. Preparation of HMW DNA from Blood Without the Use of Solvents or Enzymes

1. Collect blood as in Section 3.2., step 1.
2. Transfer 5 mL of blood to a 15-mL plastic centrifuge tube. Add 5 mL of low salt buffer and 125 µL of Nonidet P-40. Mix by gentle inversion to dissolve cell membrane.

3. Centrifuge for 10 min at 1000*g* at room temperature to recover nuclei.
4. Discard the supernatant and wash the nuclei by gently resuspending them in 5 mL low salt buffer and centrifuging again.
5. Gently resuspend the nuclei in 0.8 mL high salt buffer and transfer to a 1.5-mL microcentrifuge tube. Add 50 µl of 10% SDS and mix thoroughly by inversion. Incubate for 10 min at 55°C.
6. Add 0.3 mL of 6*M* NaCl and mix by gentle inversion. Centrifuge for 5 min at maximum speed in a microcentrifuge at 4°C.
7. Recover the supernatant and add 2 vol of absolute ethanol at room temperature. Mix by gentle inversion. Recover the DNA as in Section 3.1., step 9b, transferring finally to 1 mL of TE.
8. Check the quantity and quality of DNA as described in Section 3.1., steps 10 and 11 (*see* Notes 7, 8, and 10).

4. Notes

1. Phenol should be saturated with several changes of TE until the pH of the upper TE layer remains at 8.0. Care should be taken when handling phenol: Wear gloves, safety spectacles, and preferably work in a fume hood.
2. RNase solutions should be boiled for 5–10 min prior to use.
3. Following Proteinase K digestion, the material should be viscous before proceeding to next step. If it is not, repeat or add 100 µL more Proteinase K stock to sample and continue incubation.
4. HMW DNA should have a high viscosity because of the large size of the molecules. This will be reflected in its strandy consistency when pipeted. Always take care when pipeting and use wide bore pipets. Never vortex.
5. DNA from Section 3.1., step 9a should be of superior quality to that produced at step 9b.
6. Care must be taken not to overprecipitate or overdry the pellet at Section 3.1., step 9b as it will be very difficult or impossible to resuspend the DNA afterwards. Precipitated DNA should be hooked out when the majority of it is still relatively clear.
7. If RNA is evident when DNA is analyzed by electrophoresis (a DNase or RNase treatment can help here) use fluorimetry or Ethidium bromide staining in gels to estimate DNA concentration. If enough material is present to warrant recovery, treat with DNase-free RNase and ethanol precipitate again as in Section 3.1., step 9b.
8. If the DNA is degraded, repeat with freshly prepared RNase taking greater care with all manipulations. If problems persist, remove the RNase.

9. The yield from ~10^8 cells should be ~300 µg. The yield from 10 mL of blood should be ~100 µg by Section 3.2. and may be up to double this by Section 3.4.

10. If yields are low, check the pH of solutions and verify that the DNA has resuspended. Material should be viscous at Section 3.1., step 4 before continuing on to step 5. Repeat if necessary (*see also* Note 3).

References

1. Kornberg, A. (1980) *DNA Replication.* Freeman, San Francisco, CA.
2. Saiki, R. K., Scharf, S., Faloona, F., Mullis, K. B., Horn, G. T., Erlich, H. A., and Arnheim, N. (1985) Enzymatic amplification of ß-globin genomic sequences and restriction site analysis for diagnosis of sickle cell anaemia. *Science* **230,** 1350–1354.
3. Scharf, S., Horn, G. T., and Erlich, H. A. (1986) Direct cloning and sequence analysis of enzymatically amplified gene sequences. *Science* **233,** 1076–1078.
4. Southern, E. M. (1975) Detection of specific sequences among DNA fragments separated by gel electrophoresis. *J. Mol. Biol.* **98,** 503–517.
5. Blin, N. and Stafford, D. W. (1976) A general method for isolation of high molecular weight DNA from eukaryotes. *Nucleic Acids Res.* **3,** 2303–2308.
6. Lahiri, D. K. and Nurnberger, J. I., Jr. (1991) A rapid non-enzymatic method for the preparation of HMW DNA from blood for RFLP studies. *Nucleic Acids Res.* **19,** 5444.

CHAPTER 4

Restriction Enzyme Digestion, Gel Electrophoresis, and Vacuum Blotting of DNA to Nylon Membranes

Justin Stacey and Peter G. Isaac

1. Introduction

In 1975, Edwin Southern published a paper that revolutionized the molecular analysis of the genomes of organisms *(1)*. This procedure enabled the detection of which fragment of DNA (out of the millions that compose the genome of a higher organism) contained sequences related to that of a radioactively labeled probe. It was thus possible to detect and genetically map restriction fragment length polymorphisms (RFLPs, reviewed in ref. *2*), to know how many copies of a gene were present in an organism, to map the restriction sites around a sequence of interest, and to check if a potentially transgenic organism had really integrated the input gene into its chromosome. Suitable selection of probes, so that they detected sequences that were highly polymorphic *(3)*, enabled the identification of individuals from a population, and has led to the use of the method in forensic analysis. The identification of genes, or linked loci, affecting human disease, has also led to applications in clinical research.

Some changes have been made to the basic procedure over the years. The original procedure as described by Southern *(1)* used the capillary action of the transfer buffer to provide the motive force to transfer the DNA from the gel to the membrane. Vacuum blotting was first

From: *Methods in Molecular Biology, Vol. 28:*
Protocols for Nucleic Acid Analysis by Nonradioactive Probes
Edited by: P. G. Isaac Copyright ©1994 Humana Press Inc., Totowa, NJ

described by Peferoen et al. *(4)*, for the transfer of proteins to membranes. The vacuum blotting system described in this chapter uses a weak vacuum to blot the DNA from the gel to the membrane. Vacuum transfer has the advantage that gels can be processed relatively quickly (the original capillary procedure required overnight transfer).

In the original procedure nitrocellulose was used to capture the DNA fragments transferred from the gel. This membrane is particularly brittle, and also difficult to recycle in order to use another probe on the same samples. This has led to the development of stronger membranes that are specifically designed to be used in the blotting technique—these are primarily nylon based membranes (e.g., Pall Biodyne B, Amersham Hybond N, N$^+$, and Boehringer positively charged nylon) or polyvinylidene fluoride (PVDF) membranes (e.g., Immobilon N). In our experience the charged nylon membranes (Pall Biodyne B, Hybond N$^+$, and Boehringer positively charged nylon) work most satisfactorily with the nonradioactive probing procedures.

Wahl et al. modified the gel pretreatment procedure *(5)*, including a depurination step. The depurination breaks large fragments of DNA into smaller pieces, and this facilitates their transfer by capillary action. However, we have not found depurination to be necessary when transferring gels using the vacuum method.

Finally, various methods have been suggested for fixing the transferred DNA to the membrane. In the original procedure baking under vacuum was used. Other authors have used UV crosslinking *(6)*, or simply transferring the gel in alkaline solution *(7)*. We find that baking under vacuum gives filters with the cleanest background, and they are able to undergo many recycling steps. This is important for RFLP analysis, as several probes can be used sequentially against the same filter, saving the time and expense involved in preparing a new filter for each probe.

Until recently, the hybridization probe has always required radio-isotopic labeling (normally with ^{32}P), particularly if the genome under study was derived from a eukaryote. This placed artificial limits on the number of probes that could be manipulated by any laboratory within a set time period, and had a major impact on staff training, laboratory layout, laboratory access, and management. The recently developed nonradioactive systems (e.g., digoxigenin-labeled probe and AMPPD detection [*see* Chapter 17] or direct enzymatic labeling [e.g., *see* Chapter 20]) are now robust enough to guarantee the kind of result shown in Fig. 1, even on species with large genomes, e.g., wheat with a 2C content of 31.4–34.6 pg *(8)*. As a consequence, the method should transfer with ease to the study of smaller genomes, e.g., humans with a genome size approx one-fifth that of wheat *(9)*.

Most nonradioactive systems are particularly sensitive to background generated by careless handling of the gel or membrane. Therefore it is important that surfaces contacting the gel or membrane should be clean, and where possible the solutions should be sterile.

2. Materials

1. DNA samples: preferably standardized to the same concentration. We routinely store plant genomic DNA at 0.1 µg/µL. Store at 4°C.
2. Enzyme buffer: 10X stock. The appropriate buffer is usually supplied with the restriction enzyme when obtained from commercial manufacturers. Buffers can be made up from stock solutions, e.g., *Hin*DIII will require a 10X stock buffer of 0.5M NaCl, 0.1M Tris-HCl, pH 7.5, 0.1M MgCl$_2$, 10 mM dithiothreitol. *Eco*R1 requires a 10X buffer of 1M NaCl, 0.5M Tris-HCl, pH 7.5, 0.1M MgCl$_2$, 10 mM dithiothreitol. Store these buffers at –20°C.
3. 0.1M Spermidine: stored at –20°C (*see* Note 1).
4. Restriction enzyme: This is usually supplied in a glycerol based buffer and should be stored at –20°C. Different suppliers will supply enzymes at different concentrations. For digesting large numbers of plant genomic samples we prefer to order enzyme at a concentration of 50 U/µL.
5. Sterile deionized water.
6. Digest premix: This is made just before use (before Section 3.1., step 3). Make up enough mixture to digest all the samples. To digest a single sample of 10 µg of DNA stored at 0.1 µg/µL make 100 µL of premix: 75 µL sterile deionized water, 20 µL 10X buffer, 4 µL 0.1M spermidine, 1 µL restriction enzyme (at 50 U/µL concentration, *see* Notes 2 and 3). Mix thoroughly and store on ice until ready for use.
7. 3M Sodium acetate pH 6.0: The pH is adjusted using acetic acid, before making up to the final volume. Filter and autoclave.
8. Absolute ethanol: Store at –20°C.
9. 70% Ethanol: Store at –20°C.
10. Stretched Pasteur pipets: These are made by placing the narrow end of a pipet (held by forceps) into a Bunsen flame. When the glass begins to

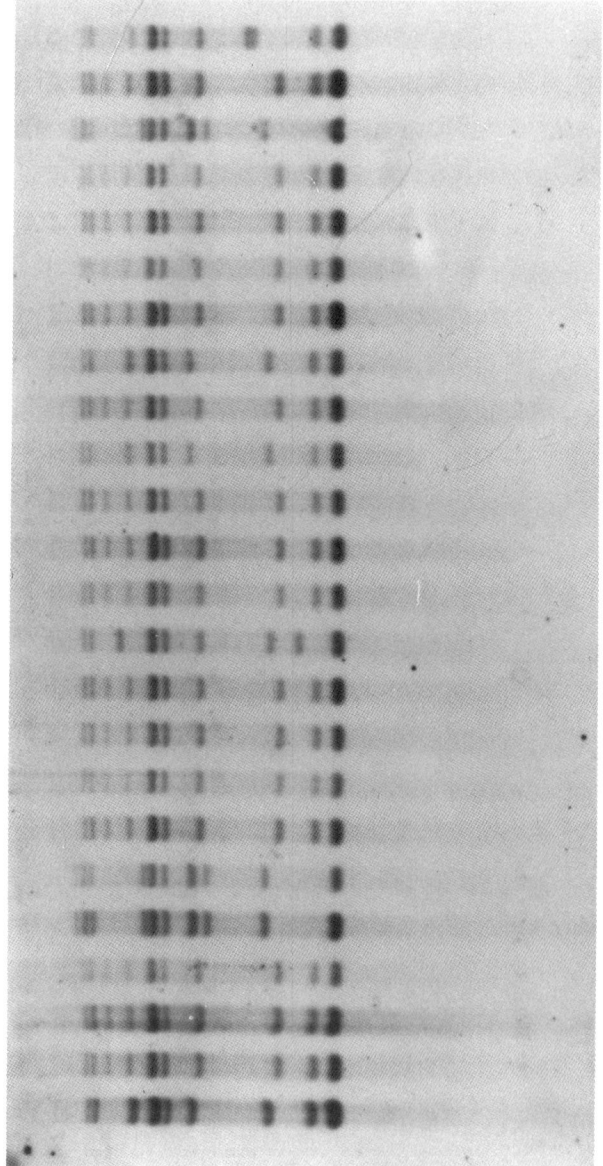

Fig. 1. A low-copy number RFLP cDNA probe used against *Hin*DIII digests of DNA from 24 barley (*Hordeum vulgare*, 2C = 10.9 pg [8]) varieties. 10 µg of digested DNA were applied to each lane of a 0.8% agarose gel that was run and transferred to Pall Biodyne B as described in this chapter. The probe was labeled with digoxigenin by the polymerase chain reaction (Chapter 10), and then hybridized against the membrane filter, and detected with AMPPD, as described in Chapter 17. The luminograph shown here is a 4-h exposure.

melt remove the pipet from the flame and stretch it with the forceps. Once it has cooled, the end of the pipet can then be broken off (wear eye protection when doing this, as small glass splinters can fly off from the broken end) to leave a fine tip.

11. TE buffer: 10 mM Tris-HCl, pH 8.0, 1 mM EDTA. Filter and autoclave.

12. Gel loading buffer stock: 80% Glycerol, 0.02% bromophenol blue, 1X TBE. Store at –20°C.

13. Loading dye: Immediately before use, mix TE and gel loading buffer stock in the ratio of 4:1, allowing 10 µL per sample.

14. Agarose (*see* Note 4).

15. 10X TBE: 0.9M Tris, 0.9M boric acid, 25 mM EDTA. Filter and autoclave.

16. Ethidium bromide: 10 mg/mL in sterile deionized water (ethidium bromide is highly toxic; always wear protective gloves when handling).

17. Gel running buffer: 2 L of 1X TBE (or appropriate amount for gel apparatus).

18. Electrophoresis apparatus (e.g., Pharmacia [Uppsala, Sweden] GNA 200 or similar) and power pack.

19. Plastic electrical insulation tape.

20. Mol-wt markers: e.g., *Hin*DIII digested bacteriophage λ DNA or prelabeled BRL 1-kb ladder (*see* Chapter 17). If digested bacteriophage λ DNA is used as a marker it should be heated to 65°C for 5 min then stored on ice until ready for use. This pretreatment dissociates the annealed "sticky-ends" of the lambda genome.

21. UV transilluminator and face shield: A full face shield must be worn when using transilluminators, as exposure to UV light may cause blindness and skin cancers.

22. Denaturing solution: 1.5M NaCl, 0.5M NaOH. Filter and autoclave.

23. Transfer solution: 20X SSC (3M NaCl, 0.3M trisodium citrate), adjust to pH 7 with HCl. Filter and autoclave.

24. Membrane prewetting solution: 2X SSC. Filter and autoclave.

25. Vacuum transfer apparatus (e.g., Millipore Milliblot V or LKB Vacugene 2016) plus adjustable vacuum pump (e.g., Millipore VacI or LKB 2016 Vacugene pump).

26. Whatman 3MM paper: Cut a piece of Whatman 3MM paper to the same size as the porous support in the vacuum blotting apparatus (for the Millipore Milliblot this is 25.5 × 25.5 cm). The porous support can be used as a template and a pencil used to draw around it. The 3MM paper can then be cut out with a clean pair of scissors.

27. Polythene mask: Cut a piece of polythene equal in size to the dimensions of the lid of the apparatus (Millipore Milliblot 32 × 32 cm). Use the lid as a template and trace around the outside with a permanent marker pen. This can then be cut out with a clean pair of scissors. Cut a window in the mask just smaller (0.5 cm in each direction) than the size of the gel to be transferred (in this way the gel should overlap the mask yet all the wells should be within the window). Use a clean ruler to measure the dimensions and draw along the ruler with a permanent marker pen. Cut out the window with a clean pair of scissors.

28. Nylon membrane: Biodyne B or similar (*see* Notes 5 and 6). Carefully cut a piece of membrane (e.g., Pall Biodyne B) 2 mm larger in each dimension than the window in the mask. Measure the dimensions of the mask window, marking the corners on the membrane with a soft pencil. Carefully cut out the membrane by scoring a sterile scalpel along a clean ruler.

29. Molten 0.8% agarose in water (about 10 mL are required). This should be made up directly before use (Section 3.3., step 8).

3. Methods
3.1. Enzyme Digestion

1. Remove the buffer and the spermidine from the freezer and defrost at room temperature. Once defrosted, vortex thoroughly and place on ice.
2. Place 100 µL of each DNA solution (10 µg) into sterile Eppendorf tubes.
3. Add 100 µL of digest premix to each 100 µL of DNA solution (*see* Note 7).
4. Mix well by flicking the tubes several times and centrifuge for a few seconds to bring all the liquid to the bottom of the tube.
5. Incubate the samples in a water bath at 37°C for at least 1 h (2 h is normal). *See* Notes 8 and 9.
6. Remove the samples from the water bath and add 0.1 vol (20 µL) of 3*M* sodium acetate, pH 6.0 followed by 2 vol (400 µL) of cold absolute ethanol.
7. Mix the samples well and place at –20°C for 2 h or at –80°C for 30 min.
8. Spin the samples in a microcentrifuge at 16,400*g* for 10 min.
9. Remove the supernatants using a stretched Pasteur pipet and discard.
10. Wash the pellets with 200 µL of 70% ethanol. Mix gently by inverting the tubes once.
11. Spin the samples in a microcentrifuge at 16,400*g* for 10 min.
12. Remove the supernatants using a stretched Pasteur pipet and discard.
13. Respin the tubes for a few seconds to collect any remaining supernatant, and remove using a stretched Pasteur pipet.

14. Resuspend the pellet in 10 μL of loading dye per 10 μg DNA. Heating the samples to 65°C for a few minutes often helps the DNA to dissolve. Mix the samples thoroughly.
15. Spin the samples for a few seconds in a microcentrifuge to collect the liquid at the bottom of the tube.

3.2. Gel Electrophoresis

This section describes the production of a 0.5-cm thick, 20 × 20 cm (i.e., 200 mL) 0.8% agarose gel in TBE, suitable for a Pharmacia GNA-200 tank. The quantities of materials used should be adjusted for other gel systems.

1. Weigh out 1.6 g of agarose and transfer to a 500-mL glass conical flask.
2. Add 180 mL of sterile deionized water and 20 mL of sterile 10X TBE.
3. Seal the conical flask with cling film and pierce in several places.
4. Microwave on medium/high setting for 2.5 min (*see* Note 10).
5. Mix gently and microwave for a further 2.5 min.
6. Mix gently and microwave for a further 1 min.
7. Mix gently (**caution:** the liquid may bump if mixed too vigorously). At this point the agarose should have completely dissolved.
8. Allow the solution to cool to 45–55°C with occasional gentle mixing (*see* Note 11).
9. While the molten agarose is cooling, seal the open ends of a clean gel casting tray with electrical insulation tape.
10. Slowly pour the cooled agarose into the gel tray.
11. Remove any air bubbles by trapping them in an inverted pipet tip.
12. Add one or more 30-place gel combs (*see* Note 12).
13. Leave the gel to set for 1 h.
14. Pour the electrophoresis buffer into the electrophoresis tank.
15. Once the agarose has solidified remove the tape from the casting tray and carefully seat the casting tray in the electrophoresis tank. Remove the comb(s).
16. Using a P20 Gilson (or similar) with a yellow tip, draw up 10 μL of the DNA solution.
17. Place the tip under the surface of the running buffer and above the opening of the well.
18. Expel the solution slowly, allowing it to sink into the well (*see* Note 13).
19. Repeat steps 16–18 for each DNA sample, loading each sample with a new pipet tip into a new well.
20. DNA mol-wt markers should be added to each gel if band sizing is to be performed. Add 10 μL (1 μg) of a diluted stock of heat treated *Hin*DIII digested λ DNA in loading buffer (or other appropriate marker, *see* Note 14).

21. When all the samples have been loaded, place the lid on the gel tank and connect the electrodes to the power pack ensuring that they are connected to the correct terminals (DNA travels toward the positive—usually red—terminal) (*see* Note 15).
22. Switch on the power pack and run the gel at 100 V for 3 h or 20 V overnight (*see* Note 16).
23. When the DNA has migrated the required distance (the dye front is normally allowed to migrate 2/3 to 3/4 of the available separation distance) switch off the power pack.
24. Remove the lid and place the gel tray in a similarly sized container such as a sandwich box or a seed tray.
25. Add 50 µL of ethidium bromide to the running buffer.
26. Pour the running buffer over the gel tray until the gel is completely submerged.
27. Shake the gel gently for 20 min.
28. Pour the running buffer into a beaker. Remove the gel and gel tray and place them on a UV transilluminator (*see* Note 17). DNA stained with ethidium bromide fluoresces orange under UV light. Photograph as required (*see* Note 18).

3.3. Vacuum Blotting

This procedure is designed for use with a Millipore Milliblot V apparatus or similar. Gels can also be transferred without special apparatus (*see* Note 19).

1. Add denaturing solution to the gel tray until the gel is completely submerged. Shake gently for 30 min.
2. Assemble the apparatus by placing the perspex "ribbed" support in the base of the apparatus. Place the porous support on top and align the 3MM paper on the porous support.
3. Wet the 3MM paper with transfer solution, and smooth out with a clean glass rod or fish slice to remove any air bubbles trapped between the paper and the support.
4. Place the mask over the 3MM paper.
5. Prewet the membrane in 2X SSC for 2 min (*see* Note 20) and place within the window in the mask aligning one edge of the membrane with the edge of the window of the mask.
6. Slide the gel from the gel tray onto the transfer apparatus, aligning the sample wells over the aligned membrane/window edge (*see* Note 21).
7. Smooth the gel with gloved fingers to expel any air bubbles trapped between both the gel and the membrane and the membrane and the 3MM paper.

8. Using a sterile Pasteur pipet, fill in the sample wells with freshly prepared molten 0.8% agarose. Allow to cool (*see* Note 22).

9. Carefully replace the upper section of the transfer apparatus, avoiding contact with the gel. Clamp in place using either the supplied butterfly nuts or bulldog clips (*see* Note 23).

10. Connect the outlet from the collection chamber to a glass 500-mL conical flask, and connect this to a vacuum pump using reinforced plastic tubing (*see* Note 24).

11. Switch on the pump and adjust to 50 mbar pressure. Allow the gel to form a seal over the mask (*see* Note 25).

12. Pour transfer solution gently onto the gel until the gel is completely submerged.

13. Transfer for 2 h, topping up with transfer solution as necessary (*see* Note 26).

14. After 2 h remove any remaining transfer solution and switch off the pump.

15. Remove the upper half of the blotting apparatus, and remove the gel to a sandwich box.

16. Remove the mask. Carefully place the membrane in 2X SSC and leave on the bench for 5 min.

17. Place the membrane DNA-side up on a piece of Whatman 3MM paper and allow to air dry (*see* Note 27).

18. Bake the membrane under vacuum at 80°C in a vacuum oven for 2 h, or UV-fix using a crosslinker, e.g., a Stratalinker (*see* Note 28).

19. Seal the membrane in a polythene bag and store at –20°C until ready for use.

20. Pour the running buffer containing ethidium bromide (saved previously, Section 3.2., step 28) over the gel until the gel is completely submerged. Shake the gel gently for 20 min.

21. View the gel on a transilluminator under UV light. If efficient transfer has taken place there should be no DNA visible (*see* Note 29).

4. Notes

1. Spermidine is used primarily in the digestion of plant or other heavily contaminated samples. It is not necessary for animal or other clean DNA.

2. The enzyme should only be removed from the freezer for a minimal time in order to preserve the activity of the enzyme.

3. Although by definition 1 U of enzyme will digest 1 μg of DNA in 1 h at 37°C, it is usual to have the enzyme in excess to ensure complete DNA digestion.

4. The brand of agarose used appears to be important as some types give high backgrounds on membranes. NBS Biologicals molecular biology grade agarose works well and gives a low background with digoxigenin labeled probes and AMPPD chemiluminescent detection.

5. There are many manufacturers of membrane and many membranes available. Positively charged membranes (e.g., Pall Biodyne B, Amersham Hybond N$^+$, Boehringer positively charged nylon) work well but are expensive. Standard nylon membranes are also suitable (e.g., Hybond N) and are often cheaper but lack the ultimate sensitivity achieved by positively charged membranes. They are sufficient for some purposes, e.g., RNA blotting (Chapters 7, 8, and 19). Polyvinylidene fluoride (PVDF, e.g., Immobilon N) membranes suffer from very high backgrounds when hybridized with digoxigenin-labeled probes and AMPPD detection, and as such are not suitable for some of the procedures in this book.

6. Membranes should be handled with great care. Use only clean coverslip forceps to handle the membrane. Never touch the membrane with fingers as this will lead to background problems.

7. If only one sample is to be digested the reagents can be added separately to the DNA tube in the order: deionized water, 10X buffer, 0.1M spermidine, enzyme.

8. Some restriction enzymes have different temperature optima, e.g., *Taq*1 has an optimum of 65°C, whereas *Bcl*1 has an optimum of 50°C.

9. If carrying out digests for a prolonged time, e.g., overnight, it is usually sensible to warm the mixture to 37°C in a water bath first, and then, after 10 min, transfer the digest to a 37°C incubator. This prevents evaporation of the contents and condensation under the lid of the centrifuge tube.

10. Alternatively the agarose can be dissolved by covering the flask with aluminum foil instead of cling film and placing in an autoclave or a pressure cooker at 15 psi for 5 min.

11. If the temperature is too high the perspex gel casting tray may warp, damaging the tray and causing uneven gel thickness; if the temperature is too cool the agarose may begin to solidify before the gel is poured.

12. Gel combs are available with varying numbers and sizes of sample wells. The procedure as described uses a 30-place Pharmacia comb with a 10-µL vol capacity. One or more combs may be used to give an increased sample throughput at the expense of separation distance.

13. Contact between the pipet tip and the sample well should be avoided as this can damage the well and lead to band distortion. An unsteady hand can result in the sample missing the well altogether, a particular problem when loading very small sample wells.

14. *Hin*DIII digested λ DNA produces band sizes of 23.13, 9.416, 6.557, 4.361, 2.322, 2.027, 0.564, and 0.125 kbp. There is a range of markers available covering different ranges of mol wts. Markers can also be prelabeled (Chapter 17).
15. It is useful to color code the electrophoresis electrodes and the power pack terminals, i.e., blue electrode to blue (negative) terminal and red electrode to red (positive) terminal.
16. The voltage level and running time will be dependent on how far the DNA samples can run in the gel. With multiple combs the running time and/or the voltage level will be lower.
17. It is not advisable to remove the gel from the gel tray as the gel can pick up contaminating material from the transilluminator. This is a common cause of background with nonradioactive probing.
18. Digested plant DNA runs as a smear from about 20 kbp downwards, with chloroplast bands visible within the smear.
19. Gels can also be transferred by capillary blotting by a slight modification of the procedure described in Chapter 7. Pretreat the gel with denaturing solution as described above (Section 3.3., step 1) and prewet the membrane (Section 3.3., step 5), then carry out steps 7–12 of the Methods section of Chapter 7, but using 20X SSC as the transfer buffer in place of 10X SSC. Continue from the rinsing step above (Section 3.3., step 16).
20. Biodyne B positively charged nylon will not wet uniformly in high salt (20X SSC) solution so use low salt (2X SSC) solution. This may not be necessary for other types of membrane.
21. Care is needed as the gels are very slippery and are easily broken. Placing gloved fingers over the open ends of the gel tray prevents the gel from slipping off the tray prematurely. A broken gel can be pieced together again on the gel tray and glued together using molten 0.8% agarose.
22. Because the gel is thinner in the wells, if they are left unfilled the transfer solution may pass preferentially through this part of the gel as it is the path of least resistance. This can cause uneven transfer of the DNA.
23. The Millipore Milliblot butterfly nuts can cause damage to the apparatus if tightened too much. Bulldog clips create a better seal and cause less damage.
24. The conical flask collects any excess transfer solution from the blotting apparatus, preventing it from reaching and possibly damaging the pump. An in-line filter (such as Millipore Millex FG$_{50}$) can be inserted between the conical flask and the pump as a further protective measure.
25. The vacuum pressure is critical; if the pressure is too high the gel will collapse, trapping the DNA in the matrix and leading to inefficient trans-

fer; if the pressure is too low the vacuum may be insufficient to hold the gel in place resulting in the gel rising to the surface of the transfer solution.

26. The gel should not be allowed to dry out as this will adversely affect the transfer. It is advisable to check the level of transfer solution every 30 min to ensure that the gel remains submerged.

27. It is best to leave the membrane on the 3MM paper until any liquid visible on the membrane has disappeared. This usually only takes 5 min.

28. Most crosslinkers have a preset optimum that should be used, e.g., Stratagene recommend 0.12 J/cm^2. A transilluminator can also be used, with a 5-min exposure to the DNA side of the membrane normally being sufficient. However, the optimal exposure does vary between transilluminators and with the age of the fluorescent tubes. It is recommended that a series of test exposures be carried out to find the optimum for critical experiments.

29. The presence of DNA in the gel implies that transfer efficiency was not optimal and is usually caused either by gel collapse (vacuum too high) or a leak/poor seal (vacuum too low).

References

1. Southern, E. M. (1975) Detection of specific sequences among DNA fragments separated by gel electrophoresis. *J. Mol. Biol.* **98,** 503–517.
2. Tanksley, S. D., Young, N. D., Paterson, A. H., and Bonierbale, M. W. (1989) RFLP mapping in plant breeding: New tools for an old science. *Biotechnology* **7,** 257–264.
3. Jeffreys, A.,Wilson, V., and Thein, S. L. (1985) Hypervariable 'minisatellite' regions in human DNA. *Nature* **314,** 67–73.
4. Peferoen, M, Huybrechts, R., and De Loof, A. (1982) Vacuum blotting: a new simple and efficient transfer of proteins from sodium dodecyl sulfate-polyacrylamide gels to nitrocellulose. *FEBS Lett.* **145,** 369–372.
5. Wahl, G. M., Stern, M., and Stark, G. R. (1979) Efficient transfer of large DNA fragments from agarose gels to DBM paper and rapid hybridization using dextran sulphate. *Proc. Natl. Acad. Sci. USA* **76,** 3683–3687.
6. Khandjian, E. W. (1987) Optimized hybridization of DNA blotted and fixed to nitrocellulose and nylon membranes. *Biotechnology* **5,** 165–167.
7. Reed, K. C. and Mann, D. A. (1985) Rapid transfer of DNA from agarose gels to nylon membranes. *Nucleic Acids Res.* **13,** 7207–7221.
8. Bennett, M. D. and Smith, J. B. (1976) Nuclear DNA amounts in angiosperms. *Phil. Trans. R. Soc. Lond.* **B274,** 227–274.
9. Kaiser, K. and Murray, N. E. (1985) The use of phage lambda replacement vectors in the construction of representative genomic DNA libraries, in *DNA Cloning: A Practical Approach,* vol. 1 (Glover, D. M., ed.), IRL, Oxford, pp. 1–47.

CHAPTER 5

Isolation of Plant RNA

Rachel Hodge

1. Introduction

Efficient extraction of high quality RNA from a variety of plant tissues is an important first step in many procedures, such as analysis of gene expression, cDNA library construction, and in vitro translation. This procedure, which is essentially as described in Draper et al. *(1),* involves grinding and phenol extraction of plant material followed by differential precipitation of RNA with sodium acetate. The protocol has been successful with leaf material and cultured cells from a large number of species, however, slight adjustments may be necessary to optimize extraction from other tissues.

The method of Hall et al. *(2)* may be appropriate if RNA is required from particularly intractable material, such as seeds. In cases where only a small amount of material is available, the micro-method of Verwoerd et al. *(3)* is highly recommended.

Successful RNA preparation depends on the inhibition of both endogenous RNase activity liberated on cell lysis and contamination of preparations by exogenous RNases. In this method treatment of equipment and solutions with diethyl pyrocarbonate (DEPC), a strong inhibitor of RNases, is used to prevent degradation of samples. In addition working quickly and keeping preparations on ice whenever possible will help to minimize problems with RNase activity.

From: *Methods in Molecular Biology, Vol. 28:*
Protocols for Nucleic Acid Analysis by Nonradioactive Probes
Edited by: P. G. Isaac Copyright ©1994 Humana Press Inc., Totowa, NJ

2. Materials

As in any procedure involving the handling of RNA it is vital to ensure that all equipment and solutions used are treated to remove ribonuclease activity and that gloves are worn at all times to prevent contamination of the preparations with ribonucleases present in perspiration. Glassware should be siliconized, soaked in 0.2% DEPC for 1 h and then baked at 180°C for 2 h. Solutions, with the exception of Tris-HCl solutions and organic solvents, should be DEPC-treated prior to autoclaving (*see* Note 1). Tris-HCl solutions should be prepared with DEPC-treated and autoclaved water. All chemicals used should be AnalaR grade.

1. Alumina Type A5 (Sigma Chemical Co., St. Louis, MO). *See* Note 2.
2. Grinding buffer: 6% 4-aminosalicylate, 1% triisopropyl napthalene sulfonate (*see* Note 3), 6% phenol (Phenol [as supplied by Fisons plc, Loughborough, UK] contains 0.1% 8-hydroxyquinoline and is equilibrated against 100 mM Tris-HCl, pH 7.5. Phenol is highly toxic; *see* Note 4 for precautions when handling.), 50 mM Tris-HCl, pH 8.4. This solution should be freshly prepared prior to use and kept on ice in the dark.
3. Phenol-resistant plastic centrifuge tubes.
4. Phenol/Chloroform: 49.95% phenol, 0.05% 8-hydroxyquinoline (equilibrated against Tris-HCl, pH 7.5 as above), 50% chloroform/isoamyl alcohol (24:1). The phenol/chloroform should be mixed in a fume-hood and left for a few hours until the mixture clears. Phenol is highly toxic; *see* Note 4 for precautions when using.
5. 4M Sodium acetate adjusted to pH 6.0 with acetic acid.
6. Absolute ethanol stored at –20°C.
7. 80% Ethanol stored at –20°C.
8. Sterile DEPC-treated water.

3. Method

As discussed above RNA preparations are vulnerable to degradation by RNases. For this reason it is important to work quickly and as far as possible keep preparations on ice at all times.

1. Harvest and weigh fresh plant material and freeze in liquid nitrogen until required. The material should be frozen as soon as possible after collection and should not be damaged or allowed to wilt.
2. Precool a pestle and mortar on ice. Grind frozen plant tissue in the presence of alumina (approx 0.5 g alumina/5 g plant material)(*see* Note 2). It is vital that the tissue is kept frozen during grinding by the addition of liquid nitrogen when required.

3. Half-fill the mortar with liquid nitrogen and before it evaporates drop in grinding buffer (2 mL/g of tissue)(*see* Note 4). Continue grinding until material and buffer are ground to a fine powder, then transfer to a 50-mL plastic centrifuge tube. Place on ice.

4. Add 1 vol phenol/chloroform and thaw on ice occasionally shaking the tube gently to homogenize contents (*see* Note 4).

5. Mix by inversion and spin at 4°C in a fixed angle rotor at 3000*g* for 10 min.

6. Remove the aqueous (top) layer being careful to avoid the protein at the interface and transfer to a fresh tube. Repeat the phenol/chloroform extraction three times or until no protein interface is visible.

7. Transfer to Corex tubes, add 0.05 vol 4*M* sodium acetate, pH 6.0 and 2.5 vol absolute ethanol that has been stored at –20°C. Mix thoroughly and spin (4°C) at 12,000*g* for 10 min to pellet nucleic acids.

8. Remove the supernatant carefully and wash the pellet with 80% ethanol, dry very briefly, and then dissolve pellet in sterile distilled water (dissolve in as small a volume as possible). It may be necessary to leave the pellet dissolving on ice for up to 1 h with occasional mixing. If the preparation still contains insoluble material spin at 12,000*g* for 5 min and discard the pellet before continuing. Depending on the volume it may be possible to transfer the dissolved nucleic acid to a sterile Eppendorf tube at this stage.

9. Add 3 vol 4*M* sodium acetate, pH 6.0, mix, and leave on ice for at least 3 h to precipitate the RNA. The preparation can be left overnight at this stage if required.

10. Spin (4°C) at 12,000*g* for 20 min to pellet the RNA. Discard the supernatant and dissolve pellet in 100–500 μL of sterile water. Reprecipitate by adding 0.05 vol 4*M* sodium acetate and 2.5 vol ethanol (–20°C), store at –20°C for 30 min and spin at 12,000*g* for 20 min. Wash the pellet with 80% ethanol, dry, and redissolve in a minimum volume of sterile water (approx 50 μL/g tissue). For long-term storage RNA should be stored as aliquots at –80°C.

11. RNA purity and concentration can be assessed by scanning spectrophotometry. Take a 1:100 dilution of RNA in sterile water and scan between 200 and 300 nm in 1-cm quartz cuvets. A clear peak of absorbance should be visible at 260 nm. An A_{260} of 1 corresponds to an RNA concentration of approx 37 μg/mL. Pure RNA has an A_{260}/A_{280} ratio of 2.0. Values significantly less than this, i.e., below 1.8, indicate contamination of the preparation, most usually with protein or phenol (*see* Notes 5 and 6). RNA quality can be checked by running on a gel (Chapter 7) (*see* Note 7).

12. Poly A$^+$ RNA can be prepared from total RNA samples as described in Chapter 6.

4. Notes

1. To DEPC treat solutions add 0.1% diethyl pyrocarbonate (DEPC), shake to disperse, and incubate at 37°C for 2 h. The solution should then be autoclaved to destroy the DEPC. DEPC is highly flammable and must be handled in a fume-hood; it is also thought to be a carcinogen so should be handled with care.
2. While weighing out alumina and grinding tissue a face mask should be worn to avoid accidental inhalation of alumina or frozen grinding buffer.
3. Both 4-aminosalicylate and triisopropyl napthalene sulfonate are toxic. A face mask should be worn while weighing them out.
4. Phenol is highly toxic and should be handled with great care and where possible confined to a fume-hood. Protective clothing, i.e., lab coat, disposable gloves, and safety glasses, should be worn at all times. If phenol, grinding buffer, or phenol/chloroform come into contact with the skin, the area should be soaked in 70% PEG 300, 30% IMS, and medical advice should be sought.
5. If protein contamination is a problem further phenol/chloroform extractions (steps 4–6) followed by ethanol precipitation (steps 7 and 8) should be carried out.
6. Phenol contamination of the preparation can be addressed by repeated ethanol precipitation (steps 7 and 8) as phenol is very soluble in ethanol.
7. When run on a gel, RNA samples should show no degradation. The two major ribosomal RNA species (28S and 18S) should be clearly visible, well defined bands (*see* Chapter 7). If the RNA appears to be degraded it indicates that exogenous and/or endogenous RNases were not sufficiently inhibited during the procedure and greater care must be taken. In practice, if the precautions set out in the protocol are adhered to no problems with RNA degradation should be experienced.

References

1. Draper, J. and Scott, R. S. (1988) The isolation of plant nucleic acids, in *Plant Genetic Transformation and Gene Expression: A Laboratory Manual* (Draper, J., Scott, R. S., Armitage, P., and Walden, R., eds.), Blackwell, Oxford, pp. 226–230.
2. Hall, T. C., Ma, Y., Buchbinder, B. U., Pyne, J. W., Sun, S. M., and Bliss, F. A. (1978) Messenger RNA for G1 protein of French bean seeds: Cell-free translation and product characterisation. *Proc. Natl. Acad. Sci. USA* **75,** 3196–3200.
3. Verwoerd, T. C., Dekker, B. M. M., and Hoekema, A. (1989) A small-scale procedure for the rapid isolation of plant RNAs. *Nucleic Acids Res.* **17,** 2362.

CHAPTER 6

Isolation of Total and Poly A⁺ RNA from Animal Cells

Ian Garner

1. Introduction

Most RNA in a mammalian cell consists of 28S, 18S, and 5S ribo-somal RNAs together with tRNAs and other small ubiquitous RNAs. The remainder (<5%) consists of messenger RNA encoding most of the polypeptides of interest to the vast majority of present day biologists. This mRNA is heterogeneous in size and generally carries long tracts of polyadenylic acid (polyA) at its 3' end. The mRNA can be purified away from other nucleic acids by hybridization of the polyA tract to oligo(dT) as described below *(1)*. Such RNA is referred to as polyA⁺ RNA and is a superior substrate for many techniques. However, purifying such molecules presents special problems as a result of the inherent instability of some RNAs and the presence of potent RNAse activities in many cell types. Suitable purification pro-tocols should include RNase inhibition or inactivation measures. The tried and tested methods described below include the latter *(2,3)*. Denaturation of all cellular proteins, including RNases, at a rate superior to that of RNA hydrolysis eliminates RNA degradation. This can be achieved using guanidinium thiocyanate and β-mercapto-ethanol which denature cellular proteins and disrupt disulfide bonds respectively. The method described below is adapted from that of Chirgwin et al. *(3),* and disrupts cells in guanidinium thiocyanate and β-mercaptoethanol prior to centrifugation through a cushion of cesium

From: *Methods in Molecular Biology, Vol. 28:*
Protocols for Nucleic Acid Analysis by Nonradioactive Probes
Edited by: P. G. Isaac Copyright ©1994 Humana Press Inc., Totowa, NJ

chloride. Under these conditions, the buoyant density of most RNAs (except small RNAs, such as tRNAs) is greater than that of other cellular components and total cellular RNA pellets at the base of the gradient. Other components, such as DNA and protein, remain in the supernatant. Many commercial kits including similar precautions are currently available and may be preferred on a speed-of-execution basis.

1.1. Ribonuclease Contamination

Whichever method of purification is chosen, other precautions to eliminate adventitious RNase contamination should also be employed, especially by the less practiced worker. Even the most experienced worker occasionally suffers from the degradation of RNA preparations associated with RNase contamination. This can be from many sources and even in some cases appear to be an "act of God." When problems are encountered, or if you are not accustomed to working with RNA, the following precautions should be employed.

All gel tanks, glassware, chemicals, and other equipment used should, if possible, be reserved for RNA preparations. Gel tanks that have been used for analysis of plasmid minipreps are often a source of RNases. Sterile disposable plasticware should be employed if possible as this is generally RNase-free. Gloves should be worn during preparation and purification procedures. These should be changed frequently.

Diethyl pyrocarbonate (DEPC) is a potent inhibitor of RNases (*see* Note 1) and is used to treat all solutions except those containing Tris. Tris solutions should be made from a reserved stock of Tris crystals with DEPC-treated sterile distilled water in suitable containers and autoclaved before use.

Glassware and other suitable equipment should be baked at 180°C for at least 8 h. If desired, glassware can be acid-washed prior to baking. Other equipment, e.g., gel tanks, can be incubated for 1–2 h at 37°C in 0.1% DEPC and rinsed with copious amounts of sterile, DEPC-treated distilled water

2. Materials

2.1. Preparation of Total RNA from Solid Tissues

1. $5M$ Guanidinium thiocyanate in $0.05M$ Tris-HCl, pH 7.5. Warm to dissolve, filter, and check pH. Conserve in shaded bottle at 4°C for <1 mo.
2. β-mercaptoethanol.

3. Antifoam A (30% [v/v]) from Sigma (St. Louis, MO).
4. Sarcosyl 30%.
5. Cesium chloride solution: 5.7M CsCl, 0.1M EDTA, 0.1% DEPC filter (0.45 μm) and autoclave. Store at 4°C .
6. Beckman (Fullerton, CA) Ultracentrifuge and SW50.1 rotor or similar.
7. 3M Sodium acetate adjusted to pH 6.0 with acetic acid.
8. Absolute ethanol (stored at –20°C in a flashproof freezer).
9. 70% Ethanol (stored at –20°C in a flashproof freezer).
10. Sterile distilled water (*see* Note 1).
11. Diethylpyrocarbonate (DEPC) (*see* Section 1.1. and Note 1 for use in treating solutions).
12. TE: 10 mM Tris-HCl, pH 8.0, 1 mM EDTA.
13. Phenol equilibrated with TE (*see* Note 2).
14. Homogenization solution: This should be made up just before use, and contains 21 mL of guanidinium thiocyanate, 0.36 mL 30% sarcosyl, 0.15 mL β-mercaptoethanol, 0.15 mL 30% antifoam A.

2.2. Preparation of RNA from Cells in Culture

Materials as in Section 2.1. and Ca^{2+}/Mg^{2+} free phosphate buffered saline (PBS) from Gibco (Grand Island, NY) or similar supplier.

2.3. Selection of Poly A$^+$ RNA Using Oligo dT

Materials 7–12 in Section 2.1. and:

1. Oligo(dT)-cellulose.
2. Plastic 1-mL syringes plugged with DEPC-treated glass wool.
3. 2X Column loading buffer: 40 mM Tris-HCl, pH 7.6, 1M NaCl, 2 mM EDTA, 0.2% SDS.
4. Column wash buffer: 1X column loading buffer.
5. Elution buffer: 10 mM Tris-HCl, pH 7.5, 1 mM EDTA, 0.05% SDS.

3. Methods

3.1. Preparation of Total RNA from Solid Tissues

1. Freshly excised tissue should be frozen in liquid nitrogen and stored at –70°C prior to use or used immediately. Using a Waring blender, or similar, that has been rinsed in DEPC-treated distilled water, break the tissue into small pieces in 6 mL of homogenization solution. Rinse the apparatus with 3 mL of homogenization solution and add to the 6 mL homogenate. Rinse the apparatus in DEPC-treated distilled water between each sample (*see* Note 3).

2. Prepare three 2-mL cushions of cesium chloride solution in polypropylene tubes (*see* Note 4) for an SW50.1 rotor. Layer 3 mL of the homogenate over each gradient.

3. Centrifuge overnight at ~150,000g (36,000 rpm in a Beckmann SW50.1) at 15°C.

4. Carefully remove the upper levels of the gradient to leave a small volume at the base of the tube. Cut away the upper portion of the tube to eliminate possible contamination. Drain the tubes by inverting the bases on paper tissues for several minutes.

5. The RNA samples may now be visible as clear pellets at the base of each tube. Recover these by the addition of 200 μL guanidinium thiocyanate solution and repeated pipeting. Each tube should then be washed with a further 200 μL guanidinium thiocyanate solution and the total 400 μL transferred to a 1.5-mL microcentrifuge tube.

6. Add sodium acetate to 0.3M and 2 vol of absolute ethanol and precipitate the RNA at –20°C overnight or –70°C for 15–30 min.

7. Recover the RNA by centrifugation (maximum speed in a microcentrifuge at 4°C for 10 min). Decant the supernatant and resuspend the RNA in 400 μL of guanidinium thiocyanate solution. Transfer to a new microcentrifuge tube.

8. Reprecipitate the RNA twice as in steps 6 and 7, finally washing the RNA pellet with 1 mL of 70% ethanol (–20°C). Dry the pellet for 1–2 min in a SpeedVac or similar. Resuspend the RNA in 200 μL of sterile DEPC-treated distilled water.

9. The absorbance of the RNA at 260 nm and 280 nm should be measured in quartz cuvets. The 260/280 ratio should be equal or close to 2.0. Significant contamination with protein will result in 260/280 ratio lower than this. If the ratio is below 1.9, extract the samples with equilibrated phenol and recover by ethanol precipitation as in 6–8. An A_{260} of 1 is equivalent to a RNA concentration to a ~37 μg/mL. Store the RNA solution at –70°C or ethanol precipitate aliquots at –70°C to be recovered as desired (*see* Note 5).

10. The material should be examined by electrophoresis through a 0.8–1% agarose gel. It should have the appearance of a smear migrating somewhere between 1 and 10 kbp DNA markers. 28S and 18S rRNA bands may be pronounced within this. If desired, denaturing polyacrylamide or agarose gels *(4)* may also be used for this purpose (*see* Note 6).

3.2. Preparation of RNA from Cells in Culture

1. Cells in suspension (10^7 to 10^9) should be recovered by centrifugation at 500g in 15-mL plastic centrifuge tubes for 10 min at room tempera-

ture. Cells grown as monolayers can be processed in culture dishes and should be rinsed in PBS to remove traces of medium.

2. Cover the cell pellet or monolayer with 9 mL of homogenization buffer. Pipet gently with a wide-bored pipet to resuspend cell pellets. Recover monolayers of cells by scraping with a rubber policeman. Cells will lyse almost immediately. Continue as from Section 3.1., step 2 (*see* Note 7).

3.3. Selection of Poly A⁺ RNA Using Oligo dT

1. Suspend the oligo(dT)-cellulose in $0.1M$ NaOH and pour into a plastic 1-mL syringe plugged with glass wool. A column of 0.1 mL will be satisfactory to process up to 1 mg of total RNA.
2. Wash the column with 1 mL sterile distilled water and equilibrate with column wash buffer until the eluate is pH 7.6.
3. Add an equal volume of 2X column loading buffer to the RNA sample to be processed.
4. Heat the sample to 65°C for 5 min to disrupt any secondary structures that may involve the poly A tail. Apply to the column and wash with 1 column vol of column wash buffer. Collect the eluate in a sterile microcentrifuge tube.
5. Repeat step 4 with the eluate.
6. Wash the column with 10 vol of column wash buffer.
7. Elute the polyA⁺ RNA with 2 column vol of elution buffer.
8. The eluate will still contain a high proportion of polyA⁻ RNA. Adjust the concentration of NaCl in the eluted sample to $0.5M$ and repeat steps 4–7.
9. Elute the polyA⁺ RNA with 2 column vol of elution buffer into a microcentrifuge tube. Add sodium acetate to $0.3M$ and 2.5 vol absolute ethanol. Precipitate the RNA at –70°C for 15–30 min.
10. Recover the RNA by centrifugation (maximum speed in a micro-centrifuge, 4°C, 10 min). Carefully decant the supernatant and wash the pellet with 1 mL of 70% ethanol prechilled to –20°C. Centrifuge as above for 1 min and carefully decant the supernatant.
11. Dry the RNA pellet and proceed as described in Section 3.1., step 8 (*see* Notes 8–10)

4. Notes

1. DEPC is harmful and extra care should be taken when handling it, e.g., manipulate stock bottles in fume-hoods wearing safety glasses, gloves, and so on. DEPC should be used at 0.1% (v/v) to treat all solutions used except those containing Tris. Treatment should be at 37°C for at least

12 h followed by autoclaving to inactivate the DEPC. Following this the solution will exhibit a sweet smell. Prepare stocks of solutions especially treated with DEPC to be reserved for RNA work, e.g., distilled water.

2. Phenol should be saturated with several changes of equal volumes of TE until the upper TE phase remains at pH 8.0. Care should be taken when handling phenol; wear safety glasses, gloves, and work in a fume-hood if possible. It is now possible to purchase distilled and pre-equilibrated preparations of phenol. Use of these reduces worker exposure and should be considered if finances permit.

3. The homogenizer used should be kept as clean as possible. It should be washed extensively with DEPC-treated distilled water prior to use, between each sample, and upon completion of a batch of preparations.

4. Polycarbonate tubes frequently fail when exposed to guanidinium thiocyanate and should never be used for this procedure.

5. Yield of total RNA from tissues depends on the source material but can be up to 20 mg RNA/g.

6. If RNA preparation is degraded, suspect first the gel tank used for electrophoresis. Has it recently been used for miniprep analysis? This is a common source of RNases. Find a clean tank or DEPC treat and repeat your gel. If this is not the case, discard solutions, make up fresh ones, and repeat taking extra care.

7. Yield of total RNA from cells in culture may vary greatly depending on cell type and growth conditions. Typically, one may expect 50–500 µg total RNA from 10^7 cells.

8. Usually, 1–2% of total RNA will be recovered as polyA$^+$ RNA by the method described. Many commercial kits utilizing a variety of approaches are now available to achieve the same aim.

9. One gram of oligodT should bind ~1 mg of polyA$^+$ RNA. It may be regenerated by washing with 0.1M NaOH, sterile distilled water, and column wash buffer.

10. In some cases, rRNA binds to oligo dT columns. If contamination is experienced, it is advisable to wash the oligo dT columns between the binding and elution steps with binding buffer containing 0.1M NaCl.

References

1. Aviv, H. and Leder, P. (1972) Purification of biologically active globin messenger RNA by chromatography on oligothymidylic acid-cellulose. *Proc. Natl. Acad. Sci. USA* **69,** 1408–1412.

2. Glisin, V., Crkvenjakov, R., and Byus, C. (1974) Ribonucleic acid isolation by cesium chloride centrifugation. *Biochemistry* **13,** 2633–2637.

3. Chirgwin, J. M., Przybyla, A. E., MacDonald, R. J., and Rutter, W. J. (1979) Isolation of biologically active ribonucleic acid from sources enriched in ribonuclease. *Biochemistry* **18,** 5294–5299.
4. McMaster, G. K. and Carmichael, G. G. (1977) Analysis of single- and double-stranded nucleic acids on polyacrylamide and agarose gels by using glyoxal and acridine orange. *Proc. Natl. Acad. Sci. USA* **74,** 4835–4838.

CHAPTER 7

Preparation of RNA Gel Blots

Rachel Hodge

1. Introduction

RNA gel blots (often referred to as Northern gel blots) are frequently used in the analysis of gene expression, for example when investigating specificity of gene expression, quantification of transcription, and analysis of RNA processing. Electrophoresis of RNA through agarose gels for blotting requires complete denaturation of the RNA. A number of denaturants have been used, including glyoxal and dimethyl sulfoxide (1), methylmercuric hydroxide (2), and formaldehyde (3). The following protocol described by Fourney et al. (4), uses formaldehyde as the denaturant but at a lower concentration than previously suggested. This allows direct visualization of the RNA without the need for staining and washing of the gel. The buffer system used is MOPS, which unlike Tris-HCl is suitable for use with formaldehyde (see Note 1). The RNA is then transferred to a nylon membrane by capillary blotting for subsequent hybridization. Mol-wt standards can be used to estimate transcript size if required.

The problems associated with exogenous and endogenous RNase activity are covered in Chapters 5 and 6 and will not be discussed here.

2. Materials

It is very important to ensure that all equipment and solutions used are treated to remove ribonuclease activity and that gloves are worn at all times to prevent contamination with ribonucleases present in

From: *Methods in Molecular Biology, Vol. 28:*
Protocols for Nucleic Acid Analysis by Nonradioactive Probes
Edited by: P. G. Isaac Copyright ©1994 Humana Press Inc., Totowa, NJ

perspiration. Glassware should be siliconized, soaked in 0.2% DEPC for 1 h and then baked at 180°C for 2 h. Plastic gel trays and other materials should be soaked in 0.1% DEPC for 1 h and rinsed out with sterile DEPC-treated water before use. Solutions should be DEPC-treated prior to autoclaving (*see* Note 2). DEPC is toxic and should be handled with care (*see* Note 2). All chemicals used should be Analar grade.

1. 10X MOPS: 0.2M 3-[N-morpholino] propanesulfonic acid sodium salt, 90 mM sodium acetate, 10 mM EDTA disodium salt, pH to 7.0 with sodium hydroxide. Store in the dark at 4°C.
2. Agarose (*see* Note 3).
3. Gel tray (the ends of which should be sealed with waterproof tape), tank, and comb soaked in DEPC water (0.1% DEPC).
4. RNA at a concentration of 5–10 mg/mL (*see* Note 4).
5. RNA mol-wt markers if required.
6. RNA sample buffer: For 1 mL mix 500 μL formamide, 100 μL 10X MOPS, 150 μL 37% formaldehyde solution (37% [w/v] as supplied), 100 μL glycerol, 100 μL 1% bromophenol blue, 50 μL 1 mg/mL ethidium bromide (*see* Note 5).
7. 10X SSC: 1.5M sodium chloride, 0.15M trisodium citrate.
8. 0.05M NaOH, 2X SSC.
9. Tray.
10. Sponge.
11. Sterile DEPC-treated water.
12. Nylon membrane: e.g., Hybond N (Amersham International, Amersham, UK). *See* Note 6.
13. 3MM filter paper (Whatman, Maidstone, UK).
14. Glass plate.

3. Method

1. These quantities are suitable for a 150 mm × 100 mm × 7 mm gel, for other systems the volumes should be adjusted accordingly. Gel preparation: Add 1.3 g of agarose and 95 mL of 1X MOPS to a 250-mL flask and heat in a microwave oven to dissolve the agarose (*see* Note 7). Transfer the flask to a fume-hood, allow the agarose to cool to around 50°C, and add 5 mL formaldehyde solution (37%). Mix gently and pour into gel tray. Leave in the fume-hood until set (30 min to 1 h).
2. Sample preparation: RNA should be thawed on ice and an appropriate amount (generally 10–20 μg total RNA or 0.5–1 μg polyA$^+$ RNA per lane) transferred to a sterile Eppendorf tube. Add 4 vol of RNA sample buffer and place at 65°C for 10 min. Transfer immediately to ice for 2

1   2   3   4

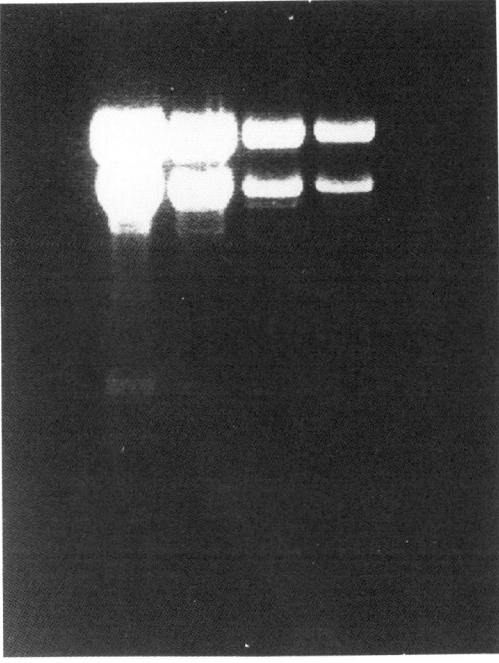

Fig. 1. Total RNA extracted from *Brassica napus* buds run on a 1.3% agarose/ formaldehyde gel as in the method. Lanes 1–4 contain 10, 5, 2, and 1 µg of total RNA, respectively.

    min, spin for 10 s in a microcentrifuge to collect the sample and place back on ice until you are ready to load the gel.

3. To set up the gel remove waterproof tape from the ends of the gel tray, place it in the gel tank, and add 1X MOPS buffer to cover the gel by 2 or 3 mm. Load the samples into the wells, connect to a powerpack, and run the gel at 50 V. For a 10-cm length gel 2–3 h should be sufficient for the dye to have migrated 75% of the length of the gel.

4. After running, photograph the gel on a UV transilluminator (302 nm), (*see* Fig. 1).

5. Transfer preparation: Soak the gel in 0.05*M* NaOH, 2X SSC for 10 min then rinse with distilled water and soak in two changes of 10X SSC for 20 min.

6. Prewet a piece of nylon membrane, cut to the same size as the gel, in sterile distilled water (DEPC-treated) and then 10X SSC (*see* Note 8).

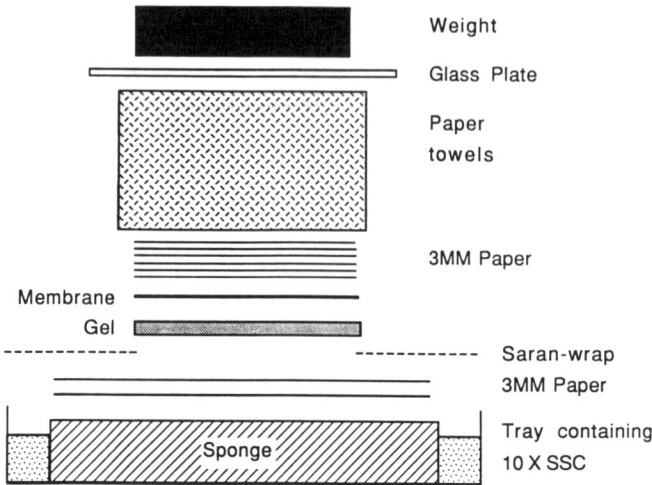

Fig. 2. Set up for blotting RNA gels.

7. Set up the blot as shown in Fig. 2. The tray should be filled with 10X SSC to 1 cm below the top of the sponge. Place two pieces of Whatman 3MM paper on the sponge and allow them to soak up the SSC.
8. Place the gel in the center of the sponge and cover the rest of the sponge and buffer tray with Saran-Wrap (Dow Chemical Co., Wilmslow, UK) so that the buffer can only move through the gel.
9. Place the prewetted membrane on the gel, carefully avoiding trapping any air bubbles between the gel and the membrane.
10. Place six pieces of 3MM paper, cut to the size of the gel and prewetted in 10X SSC, on the membrane, again avoiding air bubbles.
11. Stack around 8 cm of paper towels on the 3MM paper, followed by a glass plate and a weight, such as a 500-mL bottle filled with water (suitable for a 15 × 10 cm gel) to compress the towels. Leave overnight.
12. Carefully dismantle the blot and allow the membrane to air-dry at room temperature.
13. Wrap the filter in Saran-Wrap and crosslink the RNA to the membrane on a UV transilluminator (302 nm)—RNA side down—for approx 90 s (*see* Notes 9 and 10).
14. The membrane can now be used immediately or stored at room temperature for up to 1 mo.

4. Notes

1. Because of the reactive amine group, Tris-HCl buffers are not suitable for use when using formaldehyde as the denaturant.

2. To DEPC-treat solutions add 0.1% diethyl pyrocarbonate, shake to disperse, and incubate at 37°C for 2 h, the solution should then be autoclaved to destroy the DEPC. DEPC is highly flammable and should be handled in a fume-hood; it is also thought to be a carcinogen so should be handled with care.

3. Some types of agarose are more suitable for nonradioactive protocols than others. It may be necessary to try batches from a number of suppliers in order to optimize your system.

4. The RNA should be at a concentration that, even after the fivefold dilution with sample buffer, allows easy loading onto the gel. Appropriate volumes will depend on the gel apparatus being used, but in general it is best to keep the volumes loaded to a minimum, i.e., 10–20 μL.

5. Ethidium bromide is a carcinogen and should be handled with great care.

6. Nylon membranes that are suitable for use include both uncharged membranes, such as Hybond N (Amersham International), and positively charged membranes, such as Hybond N$^+$ (Amersham International). Nitrocellulose and polyvinylidene (PVDF) membranes are unsuitable for use with the digoxigenin/AMPPD system.

7. Although it is quite possible to dissolve agarose over a Bunsen flame it does seem that when the agarose concentration is over 1% dissolving the agarose in a microwave oven is easier owing to the tendency of thick agarose solutions to burn.

8. It is important to be extremely careful when manipulating the membrane. Avoid touching the membrane even with gloves and lift it by the corners with forceps. It is also important that the membrane wetting and transfer solutions (steps 6 and 7) are sterile.

9. After transfer to the membrane the RNA is still intercalated with ethidium bromide; this means that it is possible to mark (with a pencil) the exact positions of marker bands or lanes onto the back (not the RNA side) of the filter while UV crosslinking. This is considerably easier and more accurate than calculating the positions from a photograph.

10. The appropriate exposure time for crosslinking varies with the wavelength and age of the UV bulbs in the transilluminator. To establish the optimum exposure time a number of duplicate filters should be prepared and exposed for different lengths of time (20 s to 5 min). Subsequently filters should be hybridized, washed, and developed together, the filter giving the strongest signal indicating the optimum exposure time for crosslinking. Alternatively a UV crosslinker, such as the Stratalinker (Stratagene, La Jolla, CA), can be used according to the manufacturer's instructions.

References

1. McMaster, G. K. and Carmicheal, G. G. (1977) Analysis of single- and double-stranded nucleic acids on polyacrylamide and agarose gels by using glyoxal and acridine orange. *Proc. Natl. Acad. Sci. USA* **74,** 4835–4838.
2. Thomas, P. S. (1980) Hybridization of denatured RNA and small DNA fragments transferred to nitrocellulose. *Proc. Natl. Acad. Sci. USA* **77,** 5201–5205.
3. Rave, N., Crkvenjakov, R., and Boedtker, H. (1979) Identification of procollagen mRNAs transferred to diazobezyloxymethyl paper from formaldehyde agarose gels. *Nucleic Acids Res.* **6,** 3559–3567.
4. Fourney, R. M., Miyakoshi, J., Day, R. S., and Paterson, M. C. (1988) Northern blotting: Efficient RNA staining and transfer. *Focus* **10,** 5–7.

CHAPTER 8

Preparation of RNA Dot Blots

Rachel Hodge

1. Introduction

RNA dot hybridizations were first described by Kafatos et al. *(1)*. They allow rapid detection of transcription from a number of mRNA populations and are particularly useful in the initial characterization of clones derived from differentially expressed genes. Where accurate quantification of transcription is necessary, or many samples have to be handled, filtration manifold systems are available, such as the Millipore (Bedford, MA) MilliBlot system, that use a vacuum source to transfer nucleic acid to filter.

Although it is possible to use a pure nitrocellulose membrane matrix, the nylon based membranes currently available are much easier to handle. In addition, nitrocellulose membranes are unsuitable for use with the digoxigenin/AMPPD system described in Chapter 19. This protocol is a slightly modified version of that supplied with Hybond N (Amersham International, Amersham, UK).

The problems associated with exogenous RNase activity are discussed in Chapters 5 and 6, and will not be dealt with here.

2. Materials

All chemicals used should be AnalaR grade. Solutions should be DEPC (diethyl pyrocarbonate) treated prior to autoclaving (*see* Note 1), to destroy ribonuclease activity. Gloves should be worn at all times, particularly when handling membranes and RNA solutions.

From: *Methods in Molecular Biology, Vol. 28:*
Protocols for Nucleic Acid Analysis by Nonradioactive Probes
Edited by: P. G. Isaac Copyright ©1994 Humana Press Inc., Totowa, NJ

1. Membrane: Hybond N (Amersham International).
2. Sterile DEPC-treated water.
3. 10X SSC: 1.5M sodium chloride, 0.15M trisodium citrate.
4. 20X SSC: 3.0M sodium chloride, 0.3M trisodium citrate. Store on ice before use.
5. RNA: Total RNA can be prepared by a number of methods (e.g., Chapters 5 and 6) and should be checked for degradation by electrophoresis (*see* Chapter 7). An RNA concentration of 10 mg/mL or above is ideal (*see* Note 2).
6. 10X MOPS: 0.2M 3-[N-morpholino] propanesulfonic acid sodium salt, 90 mM sodium acetate, 10 mM EDTA disodium salt, pH to 7.0 with sodium hydroxide. Store in the dark at 4°C.
7. RNA incubation solution: To prepare 1 mL mix 657 μL formamide, 210 μL 37% formaldehyde solution (37% [w/v] as supplied) and 133 μL 10X MOPS (*see* Note 3). This solution may be prepared fresh or stored at –20°C indefinitely. Both formamide and formaldehyde are toxic and should be handled, with care, in a fume-hood.
8. Hair drier (optional).

3. Method

1. The membrane should be cut to a suitable size (allow 1 cm^2 per RNA dot) and marked (with a pencil) to show orientation and positions for sample loading (*see* Note 4).
2. Wet the membrane by laying it on the surface of distilled water and then wash briefly in 10X SSC and air-dry thoroughly.
3. Thaw the RNA samples on ice and transfer appropriate amounts (*see* Notes 5 and 6) to fresh tubes containing a 3X vol of RNA incubation solution, mix thoroughly.
4. Heat the RNA samples at 65°C for 5 min to denature RNA secondary structure, then cool on ice.
5. Add an equal volume of ice-cold 20X SSC and mix thoroughly.
6. Dot the RNA solution onto the membrane in 2-μL aliquots using a Gilson pipet, dry the membrane between each loading. To speed up the process it is possible to use a hair-dryer to dry the membrane between loadings.
7. After loading the last sample dry the membrane thoroughly, wrap in Saran-Wrap (Dow Chemical Company, Wilmslow, UK) and crosslink the RNA to the membrane on a UV transilluminator (302 nm)—RNA side down—for 90 s (*see* Note 7).
8. The membrane can then be used in hybridization experiments immediately or stored, wrapped in Saran-Wrap, at room temperature for up to 1 mo.

4. Notes

1. To DEPC-treat solutions add 0.1% diethyl pyrocarbonate, shake to disperse, and incubate at 37°C for 2 h. The solution should then be autoclaved to destroy the DEPC, which if present may carboxymethylate purine residues in the RNA. DEPC is highly flammable and should be handled in a fume-hood. It is also suspected to be a carcinogen so should be treated with respect.

2. Because of the subsequent eightfold dilution of the sample by addition of incubation solution and 10X SSC it is highly advisable to have RNA preparations at a concentration of 10 mg/mL or above. Preparations of less than 10 mg/mL should be ethanol precipitated (Chapter 5) and redissolved at higher concentrations.

3. MOPS buffer is used where formaldehyde is a component of the incubation mixture. Tris buffers are not suitable owing to the reactive amine group.

4. When handling membranes great care should be taken. Pick them up by the corners with forceps and avoid touching the membranes even while wearing gloves.

5. The amount of RNA loaded per dot is obviously dependent on the abundance of the transcript in the RNA population. For a cDNA clone identified by differential screening (and hence of high abundance in the mRNA pool) a loading of 5 μg total RNA per dot should be sufficient. For medium and low abundance transcripts it may be necessary to load more total RNA or isolate polyA$^+$ RNA for efficient detection of transcript.

6. Where duplicate filters are being prepared, master mixes of RNA, incubation solution, and 10X SSC should be made for each RNA sample.

7. The appropriate exposure time for crosslinking varies with the wavelength and age of the UV bulbs in the transilluminator. To establish the optimum exposure time a number of duplicate filters should be prepared and exposed for different lengths of time (20 s to 5 min). Subsequently filters should be hybridized, washed, and developed together. The filter giving the strongest signal indicates the optimum exposure time for crosslinking. Alternatively a UV crosslinker (e.g., a Stratalinker from Stratagene [La Jolla, CA]) can be used.

References

1. Kafatos, F. C., Jones, C. W., and Efstratiadis, A. (1979) Determination of nucleic acid sequence homologies and relative concentrations by a dot hybridization procedure. *Nucleic Acids Res.* **7**, 1541–1552.

CHAPTER 9

Isolation of Plasmids
for the Preparation of Probes

Peter G. Isaac and Justin Stacey

1. Introduction

Most molecular probes are produced by cloning, or subcloning, the required "foreign" DNA fragment (i.e., the DNA sequence that will be used as the final probe) into a bacterial plasmid. This chapter describes the basic procedures necessary to produce plasmids of sufficient purity to label directly, or to use as polymerase chain reaction (PCR) substrates (and hence amplifying the "foreign" DNA fragment). If the plasmid is to be used as a substrate for PCR amplification, then it is only necessary to follow the steps outlined in Section 3.1. This procedure was devised by Birnboim and Doly (1) and produces large yields of plasmid from bacterial cultures of the order of 1 mL, but the plasmid is contaminated with RNA (particularly the tRNA that is added as a carrier during the procedure) and with the genomic DNA of the host bacterium.

If the plasmid is to be labeled directly, without either PCR amplification of the insert or gel purification of the insert (*see* Chapter 11), then the plasmid will require purification on cesium chloride (CsCl). If this is the case the procedure described in Section 3.2. should be followed, after obtaining an initial purification of the plasmid as described in Section 3.1. If CsCl purification is to be used then it is advisable to extract DNA from larger volumes of culture (10 mL is

From: *Methods in Molecular Biology, Vol. 28:*
Protocols for Nucleic Acid Analysis by Nonradioactive Probes
Edited by: P. G. Isaac Copyright ©1994 Humana Press Inc., Totowa, NJ

usually sufficient). Isopycnic centrifugation in CsCl purifies the DNA as a band that floats in the middle of the centrifugation tube, at the point where the buoyant density of the plasmid matches the buoyant density of the CsCl medium. RNA contaminants (with the exception of small RNA oligonucleotides) form a pellet at the bottom of the centrifugation tube, whereas proteins form a pellicle that floats on the surface of the cesium chloride solution *(2)*. The plasmid DNA (if intact and supercoiled) will have a buoyant density different from that of the host chromosome DNA, and bands further down the tube than the chromosomal DNA *(2)*.

2. Materials
2.1. Birnboim-Doly Plasmid Isolation

1. $1M$ Tris-HCl, pH 8.0: Dissolve 12.1 g Tris base in 70 mL of water. Add concentrated hydrochloric acid until the pH is 8.0. Make up to a final volume of 100 mL. Filter through a paper filter and autoclave before storing the solution in a refrigerator.
2. $0.5M$ EDTA: Weigh out an appropriate amount of disodium ethylene-diaminetetraacetic acid and add to stirring distilled water (about 75% of the final volume). Add sodium hydroxide pellets slowly until the solution begins to clear. Monitor the pH and add NaOH until the EDTA has dissolved and the pH reaches between 7 and 8. Make up to the final volume, filter through a paper filter, and autoclave. Store the solution in a refrigerator.
3. 20% Glucose: Autoclave and store the solution in the fridge. Glucose solutions should not be autoclaved more than once.
4. Sterile distilled water: Filter distilled water through a paper filter, autoclave. Store at room temperature.
5. Lysis solution: This should be made from stock solutions directly before use. To make 10 mL of solution, mix 9 mL sterile distilled water, 0.25 mL $1M$ Tris-HCl, pH 8.0, 0.2 mL $0.5M$ EDTA, pH 8.0, 0.5 mL 20% glucose. Finally add 20 mg of lysozyme and keep the solution chilled on ice.
6. $5N$ NaOH: Make up with sterile water; there is no need to autoclave. Store this solution in room temperature in a plastic bottle.
7. 10% (w/v) Sodium dodecyl sulfate (SDS): Filter through a paper filter, autoclave, and store at room temperature. SDS may precipitate out if the solution becomes cold. If this happens, stand the solution in a water bath or incubator at >30°C until it clears.

8. NaOH/SDS solution: This solution should be made from stock solutions directly before use, and kept at room temperature. For 10 mL of solution, add 0.4 mL 5*N* NaOH to 8.6 mL of sterile distilled water. Mix thoroughly and then add 1 mL of 10% SDS.

9. 3*M* Sodium acetate, pH 4.9: Adjust the pH with glacial acetic acid before making up to the final volume. Filter through a paper filter and autoclave. Store this solution at room temperature.

10. 10 mg/mL Yeast tRNA: Make up 1 mL in sterile water and keep frozen at –20°C (*see* Note 1).

11. High salt solution: For 1.5 mL of solution, mix 4.5 μL of 10 mg/mL yeast tRNA solution (*see* Note 1) with 1.5 mL 3*M* sodium acetate, pH 4.9 solution. Make this solution up immediately before use and keep on ice.

12. Luria broth: 5 g yeast extract (Difco), 10 g NaCl, 10 g Bactotryptone (Difco, Detroit, MI) dissolved in 1 L of distilled water. Sterilize by autoclaving. Add 0.01 vol of sterile 20% glucose before use.

13. 100X Concentrate of appropriate antibiotic (*see* Note 2): Ampicillin is generally used at a final concentration of 100 μg/mL, and so the concentration for the 100X stock is 10 mg/mL. 100X Tetracycline is 2.5 mg/mL.

14. Ethanol (store at –20°C).

15. 3.0*M* Sodium acetate, pH 6.0: Adjust the pH with acetic acid before making to the final volume. Filter and autoclave this solution and store at room temperature.

16. 0.1*M* Sodium acetate, pH 6.0.

17. 70% Ethanol (store at –20°C).

18. TE: 10 m*M* Tris-HCl, pH 8.0, 1 m*M* EDTA. Filter and autoclave.

19. 20% Glycerol.

2.2. Cesium Chloride Purification of Plasmids

In addition to the materials listed in Section 2.1., the following are required.

1. Cesium chloride (CsCl) solid.

2. 10 mg/mL Ethidium bromide. This is a carcinogen, wear gloves while using.

3. Balancing solution: Add 10 g CsCl and 0.8 mL 10 mg/mL ethidium bromide to 9.2 mL of TE.

4. Paraffin oil.

5. Ultracentrifuge, rotor, tubes, and caps capable of forming a CsCl gradient. In the example given in this chapter a Beckman (Fullerton, CA) TL100 centrifuge, TLA-100.2 rotor, and 2.0 mL heat sealable (Quick-Seal) tubes are used (*see* Note 3).

6. Small UV light source. We use a cheap battery powered lamp "Minifluorescent Lantern: Ultra Violet Fluorescent Tube" code JML1191 available from Ultra-Violet Products (San Gabriel, CA).

7. NaCl/water saturated *n*-butanol (i.e., butan-1-ol): Make a saturated solution of NaCl by boiling NaCl crystals in water until no more dissolve. Allow the solution to cool to room temperature, pour off the aqueous layer into a bottle, and add approx 1/2 vol of butan-1-ol. Shake this mixture vigorously (salt may precipitate out; this will not cause problems). The organic material will settle as the upper layer.

8. TEN: 10 mM Tris-HCl, pH 8.0, 0.1 mM EDTA, 100 mM NaCl. Filter and autoclave.

9. Dialysis tubing (optional; *see* Section 3.2., step 6): preboiled in 1 mM EDTA and rinsed in sterile distilled water.

10. Stretched Pasteur pipets: These are made by placing the narrow end of a pipet (held by forceps) into a Bunsen flame. When the glass begins to melt remove the pipet from the flame and stretch it with the forceps. Once it has cooled, the end of the pipet can then be broken off (wear eye protection when doing this, as small glass splinters can fly off from the broken end) to leave a fine tip.

3. Methods

3.1. Birnboim-Doly Plasmid Isolation

The volumes given in this extraction procedure are designed to extract the plasmid from about 1 mL of culture. If CsCl purification is to be used (Section 3.2) then the volumes of culture and reagents should be scaled up 10-fold.

1. From a single bacterial colony on an agar plate, or a previously grown overnight culture, inoculate a small volume (e.g., 5 mL) of Luria broth supplemented with an appropriate antibiotic (*see* Note 2). Place the culture in a shaking incubator or water bath and leave to grow overnight (16–18 h) (*see* Note 4).

2. Chill the culture on ice for 5 min then transfer between 0.75 and 1 mL of the culture to a microcentrifuge tube (also on ice) and spin the cells down for 5 min at top speed in a microcentrifuge (the centrifuge can be at room temperature). Remove the supernatant into a container and autoclave it before disposal.

3. Resuspend the pellet in 100 μL of cold lysis solution. It is worthwhile making sure that all the clumps of cells are disrupted at this stage, as cells that do not break here will contribute protein and chromosomal

DNA contamination later. The best way to resuspend the cells is to blow the material in and out of a pipet tip several times and vortex the tube. The tubes can also be struck like a match on the base of a wire rack before vortexing. Leave the mixture on ice for 30 min.

4. Add 200 μL of NaOH/SDS solution to the cells. Mix this solution in gently by holding the tube horizontally and flicking it several times. Return the tube to the ice bucket for 5 min. The mixture will at first become clear and viscous, then cloudy.

5. Add 150 μL of high salt solution. Mix this in gently (as above) and then return the tube to the ice bucket for 1 h. During this stage the SDS, protein, and chromosomal DNA forms a white clotted mass, but the plasmid DNA, because of its size and covalently closed nature, remains in solution. Once the clot has started to form do not disturb the tube—ideally all the clot should remain in a single solid mass.

6. Spin out the clot for 5 min in a microcentrifuge (the centrifuge can be at room temperature).

7. Remove as much of the clear supernatant as possible (about 400 μL) into a fresh tube. If some of the clotted material contaminates the supernatant at this stage, it can be removed in step 10 below.

8. Add 2 vol (800 μL) of cold ethanol, vortex thoroughly, and leave the mixture at –20°C for 30 min (*see* Note 5).

9. Spin down the nucleic acids in a microcentrifuge at top speed for 2 min and discard the supernatant. The nucleic acids should be visible as a white pellet.

10. Redissolve the pellet (there is no need to first remove the excess ethanol under vacuum) in 100 μL of 0.1M sodium acetate, pH 6.0. This takes about 5 min at room temperature with occasional agitation. If some clotted material contaminated the supernatant in step 7 above then this material can now be removed by centrifuging the preparation for 10 min in a microcentrifuge—the supernatant should then be decanted into a fresh tube.

11. Add 2 vol of cold ethanol, vortex, and return the tubes to the –20°C freezer for 10 min (*see* Note 5).

12. Repellet the nucleic acids by centrifugation as in step 9 above. If culture volumes of more than 1 mL have been extracted it may be convenient at this point to redissolve the pellets in a small volume of 0.1M sodium acetate, pool them into a single microcentrifuge tube, and reprecipitate them (i.e., from step 11 above).

13. Dry the pellet under vacuum.

14. Dissolve the nucleic acid in TE (100 μL per original 1 mL of culture). This can be stored either at 4°C or at –20°C.

15. To use this plasmid for a PCR template, prepare a 1 in 50 dilution of the final preparation in 20% glycerol. This should be stored at –20°C until required. One microliter of this is usually sufficient for a 25-cycle amplification (e.g., *see* Chapter 10).

3.2. Cesium Chloride Purification of Plasmids

1. Measure 0.92 mL of plasmid in TE (accurately) into a microcentrifuge tube (*see* Note 6).
2. Add 1.0 g CsCl and 80 µL 5 mg/mL ethidium bromide to the tube (*see* Note 6). Cover the tube and invert a few times to dissolve the CsCl. Transfer to a heat sealable tube and top up the tube with balancing solution if necessary (*see* Notes 7 and 8).
3. Seal the centrifuge tube (either by heat sealing or adding a cap fixing) and centrifuge the samples at 435,680g (100,000 rpm) for 16 h in the TLA-100.2 (or equivalent g × h in other rotors, *see* Note 9). The rotor temperature should be maintained at 20°C. When the rotor is decelerating the last 5000 rpm should be decelerated without braking.
4. Remove the tube from the rotor. The red DNA bands are usually visible in ambient light in the middle of the tube, but it is necessary to view them under UV light to gain the full impression of their size. There are usually two bands visible—the upper band (usually fainter) is chromosomal DNA and the lower band contains supercoiled plasmid DNA. Collect the plasmid by making a small hole with a pin in the top of the tube (if the tube is of the heat seal type) or loosening the cap (if it has a metal fixing), then puncture the side of the tube with a wide-bore needle attached to a 2-mL syringe. Draw the plasmid gently into the syringe (collecting the minimum possible volume) and transfer it to a microcentrifuge tube.
5. Remove the ethidium bromide by extracting four to six times with NaCl/water saturated *n*-butanol. This is done by adding an approximately equal volume of butanol to the DNA sample and repeatedly inverting the tube. The ethidium bromide moves into the upper (organic) layer. Collect and dispose of the butanol/ethidium bromide mixture in accordance with the local waste disposal procedures.
6. Either dialyze the clear aqueous phase against 1 L of TEN for 2 h; or dilute the plasmid with 3 vol of TEN. The procedure used is determined by the number of samples and their volumes.
7. Divide the preparation into 0.4-mL aliquots in Eppendorf tubes and add 40 µL 3M sodium acetate, pH 6.0 and 1 mL of cold ethanol. Invert the tubes a few times and precipitate out the DNA at –20°C for more than 2 h.

8. Spin in a microcentrifuge for 10 min at full speed and remove all of the supernatant with a stretched Pasteur pipet.
9. Add 1 mL of cold 70% ethanol, invert the tube once, and immediately respin for 10 min.
10. Remove all of the supernatant and dry the pellets under vacuum. Add TE to the pellets (so that the total final volume is equal to the original plasmid solution volume, usually 1 mL as in step 1 above). Leave the samples overnight at 4°C to dissolve. Heat the tubes for 10 min at 65°C then mix gently and pool DNA solutions. The empty tubes can be washed with a small amount of TE that can be pooled with the main bulk.
11. The amount and integrity of a plasmid can be checked by electrophoresis (*see* Note 10 and Chapter 4) and spectrophotometry (*see* Chapter 2).

4. Notes

1. The tRNA is only necessary when extracting from 1 mL or less of culture; it is unnecessary when extracting from 10 mL cultures.
2. The antibiotic used is generally the same that was used in the original selection of transformed bacteria. For instance, for the pUC series this would be ampicillin, for pBR322 with insertions in the *Pst*1 site this would be tetracycline, and for pBR322 with insertions in the *Eco*RV or *Bam*H1 sites this would be ampicillin.
3. In general, vertical or angled rotors are used, and the speed at which they are used is normally calculated from the maximum speed of the rotor, multiplied by a derating factor calculated from the specific gravity of the contents of the tube. For CsCl gradients this is normally 0.77 *(3)*. Further derating may be applied to particular rotors, because of their age and previous history *(3)*.
4. If the plasmid to be prepared has a ColE1 derived replication origin (e.g., the pUC series or pBR322) then the culture may be supplemented with chloramphenicol after 2.5 h of growth *(4)*. However, as perfectly acceptable yields of plasmids result without chloramphenicol amplification, the use of this treatment is not recommended. We have occasionally had problems separating the plasmid DNA from the cell debris with chloramphenicol amplified cultures.
5. If this time is exceeded then the final pellet is very much more difficult to dissolve and the final solution appears more viscous.
6. It is important that the volume is measured accurately, and the CsCl crystals are also weighed accurately. Small errors in the amounts will affect the buoyant density of the final solution, displacing the plasmid band either up or down the centrifuge tube after the centrifugation step.

7. Alternatively, the tubes can be topped up with paraffin oil (although paraffin should not comprise more than half the sample). If using paraffin oil, the same relative volumes of DNA solution and paraffin oil must be used in samples placed opposite one another in the centrifuge rotor.

8. Keep the solution in the dark once the ethidium bromide has been added until the samples are ready to load in the rotor.

9. The actual time taken depends to a certain extent on the volume of the sample, the geometry of the tube, and the g force to which the sample is exposed. Small rotors, e.g., the Beckman TLA-100.2, will equilibrate in 16 h, whereas larger rotors will take longer. In general, the more vertical the orientation of the tube, the more quickly the CsCl gradient forms. The rotor can be stopped and inspected, and the centrifuge run continued without disturbing the gradients, provided that the samples are handled carefully, and the centrifuge is decelerated for the last 5000 rpm without braking.

10. It should be noted that more than one band will appear on electrophoresis of a plasmid. In general, the covalently closed supercoiled form will comprise most of the DNA, and this runs the most rapidly through an agarose gel because of its compact nature. This is then followed by linear molecules, relaxed (nicked) circular forms and dimer, trimer, and other concatenenes.

References

1. Birnboim, H. and Doly, S. (1979) A rapid alkaline extraction procedure for screening recombinant plasmid DNA. *Nucleic Acids Res.* **7,** 1513–1523.

2. Birnie, G. D. (1978) Isopycnic centrifugation in ionic media, in *Centrifugal Separations in Molecular and Cell Biology* (Birnie, G. D. and Rickwood, D., eds.), Butterworths, London, pp. 169–217.

3. Molloy, J. and Rickwood, D. (1978) Characteristics of ultracentrifuge rotors and tubes, in *Centrifugal Separations in Molecular and Cell Biology* (Birnie, G. D. and Rickwood, D., eds.), Butterworths, London, pp. 289–316.

4. Clewell, D. B. (1972) Nature of ColE1 plasmid replication in *Escherichia coli* in the presence of chloramphenicol. *J. Bacteriol.* **110,** 667–676.

CHAPTER 10

Production of Hybridization Probes by the PCR Utilizing Digoxigenin-Modified Nucleotides

Tom McCreery and Tim Helentjaris

1. Introduction

The polymerase chain reaction (PCR) *(1)* is an efficient method for copying a fragment of DNA when primers exist that flank the target sequence. We have found that since most cloning vectors utilize the *lacZ* gene with a synthetic multiple cloning site *(2)*, a single pair of primers will amplify sequences inserted into many common phage and plasmid vectors, such as the pUC plasmid and Lambda Zap phage series and their derivatives. We have had little difficulty amplifying inserts up to 3000 bp in length and now consider this as a preferred alternative to growth of bacterial cultures with subsequent nucleic acid purification to produce DNA fragments of this length for both mapping and sequencing.

Since the *Taq* polymerase also has little difficulty incorporating digoxigenin-11-dUTP into its products, we have found that PCR is also a preferred method for producing hybridization probes. It has the following advantages:

1. It utilizes very little input target DNA compared to the final yield of the product.
2. It can utilize either purified DNA (both super-coiled and linear molecules), broth culture of a plasmid-infected bacteria, or the lysate of a phage-infected bacteria culture as the source of the target sequences.

From: *Methods in Molecular Biology, Vol. 28:*
Protocols for Nucleic Acid Analysis by Nonradioactive Probes
Edited by: P. G. Isaac Copyright ©1994 Humana Press Inc., Totowa, NJ

3. The progress of the labeling reaction can be easily followed by checking the production of double-stranded product by standard agarose gel electrophoresis.

It also has the three following disadvantages:

1. The target sequences must be cloned into a vector for which flanking primers are available.
2. We have difficulty amplifying most sequences larger than 3000 bp in length.
3. Some sequences <3000 bp in length are recalcitrant to PCR-amplification, presumably because of high G-C content or the presence of snapback sequences.

Nevertheless, we find this technique to be preferable to oligo-labeling (*see* Chapter 11) and usually attempt to use it for every target sequence until it is determined to be intractable to this approach.

2. Materials

1. 10X PCR buffer: Mix 3.7 mL of sterile distilled water, 5.0 mL of 500 mM KCl, 1.0 mL of 1M Tris-HCl, pH 8.2, 0.2 mL of 1M MgCl$_2$, 0.1 mL of 1% gelatin.
2. Deoxynucleotide triphosphates (dXTPs): Make up a single solution containing 2.5 mM of dATP, dGTP, dCTP, and dTTP.
3. Digoxigenin-11-dUTP: Dilute stock from Boehringer Mannheim by adding 375 µL of distilled water to the 25 µL containing 25 nmol of modified base.
4. Primers: Primers can be specific to each individual application (i.e., usually this simply reflects the cloning vector); however, we use a single pair of primers that amplify inserts in pUC-based plasmids, which are also useful for any other vector with a *lacZ*-multiple cloning site.

    100 µM PCRFSEQ: 5' TTGTA AAACG ACGGC CAGTG 3'
    100 µM PCRRSEQ: 5' GGAAA CAGCT ATGAC CATGA T 3'

    See Note 1.
    Mix these primers 1:1 to get a 50 µM solution of each primer.
5. *Taq* DNA polymerase: Supplied by the manufacturer at a concentration of 5 U/µL.
6. 1X Reaction mix: 24 µL of this mix are required for each sample. To make 24 µL of this mix combine 19.4 µL of sterile distilled water, 2.5 µL of 10X PCR buffer, 0.5 µL of 2.5 mM dXTP, 1.0 µL of dilute digoxigenin-11-dUTP (*see* Note 2), and 0.5 µL of mixed primers.

Vortex this mixture and then add 0.1 µL of *Taq* polymerase; mix by pipeting up and down (*Taq* polymerase seems to be very sensitive to vortexing).

7. Mineral oil: We use U.S.P. grade heavy mineral oil available in drug-stores.

8. LB (Luria Bertani) broth: Combine 10 g of Tryptone, 5 g of yeast extract, and 10 g of NaCl in 1 L of H_2O. Sterilize by autoclaving.

3. Method

PCR conditions will vary depending on the primer pairs utilized. Since we have found that use of higher annealing temperatures can facilitate the amplification of sequences with high G-C content, the development of primers with an annealing temperature of at least 60°C and higher is recommended.

A PCR cycle consists of three steps, The first is a denaturation step that separates the double-stranded DNA into its two single-stranded components. The second is an annealing step where the primers and single-stranded target DNAs anneal to each other. The third step is the actual amplification of the input where the *Taq* polymerase copies the target DNA. A sample PCR program would contain 1 cycle of 95°C for 2 min, 60°C for 30 s, and 72°C for 3 min. This cycle has a longer time at 95°C to provide the initial denaturation of the target DNA. This would be followed by 18 cycles of 95°C for 30 s, 60°C for 30 s, and 72°C for 3 min. The last cycle is 95°C for 30 s, 60°C for 30 s, and 72°C for 10 min. This provides for a longer amplification step. Some thermocyclers can be programmed to hold at 6°C for 18 h to allow for overnight runs.

1. 1 µL of an overnight broth culture or lambda phage lysate (*see* Note 3) or 1–10 ng of target DNA (*see* Note 4) is added to a 0.65-mL microcentrifuge tube or to a single well in a 96-well microtiter plate (for those thermocyclers capable of utilizing them).

2. Prepare the reaction mix in bulk for all of the reactions to be performed and then aliquot 24 µL of the reaction mix into each tube. Mix by pipeting up and down

3. After the mixture is added to the tubes, add two drops of mineral oil to prevent evaporation during amplification.

4. Place the tubes, or the microtiter plate, into the thermocycler and start the program.

5. After the last 72°C extension step is complete, the samples can be removed at any time or they can be allowed to remain overnight at 6°C or 25°C depending on the thermocycler. After removing the tubes from the thermocycler, draw off the bottom aqueous layer with a micropipet and transfer to a fresh tube.
6. The expected yield for this reaction is approx 500–1000 ng depending on the size of the target fragment. A reasonably accurate measure of the yield of this reaction can be obtained by checking 4 μL of the reaction products on a small agarose gel. The reaction can be scaled up quite easily to obtain more product if necessary. The probes produced by this method are stable for several months or more at –20°C (*see* Note 5).

4. Notes

1. Primer selection: By using a longer primer you can raise the annealing temperature in the PCR reaction. We have experimented with other primers for the *lacZ*-MCS vectors and have found that the following pair of primers also amplifies efficiently and allows the use of even higher annealing temperatures (to 68°C). This can further increase the percentage of sequences that are amplified by PCR.

   FUPSTRM:5' ATGTG CTGCA AGGCG ATTAA GTTGG G 3'
   RDNSTRM:5' CACAC AGGAA ACAGC TATGA CCATG 3'

2. Proportion of modified base: The protocol described above calls for 5% labeled base. This provides for a strong signal with minimal background on most autoradiograms. Concentrations as low as 2.5% and as high as 20% are usable although we recommend 5% for the best overall results (*see* Fig. 1).
3. Lambda phage lysates. These can be labeled directly by PCR by using the following protocol. Prepare plaques in the normal manner except that both top and bottom layers in the plate are made with agarose, not agar, as impurities in the agar will inhibit the PCR reaction. Pick individual plaques and transfer them either to 0.65 mL microcentrifuge tubes or 96-well plates. Add 100 μL of LB broth to each plaque and grow overnight at 37°C with shaking. Add 1 μL of this lysate to PCR tubes or plates. Continue with standard protocol
4. Target DNA concentration. This is crucial to both a good amplification or labeling reaction. This is also a case where more is definitely not better. Keeping the concentration of target DNA close to 1 ng will maximize the yield of a labeled probe with higher specific activity.

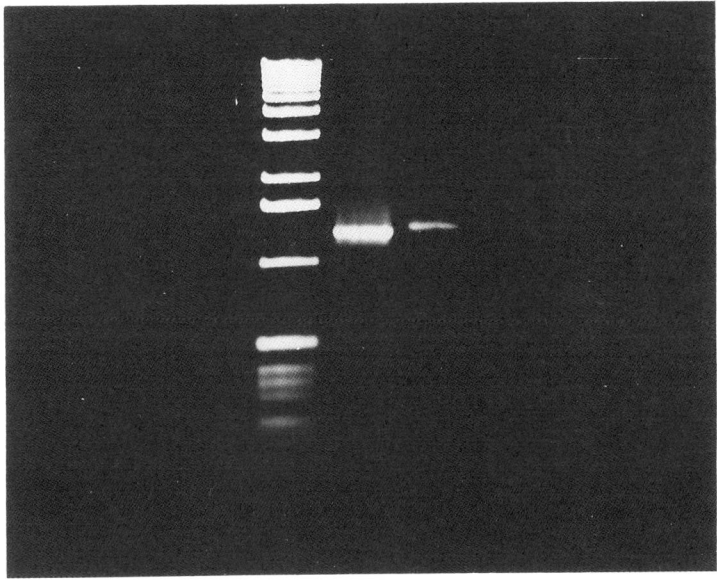

Fig. 1. The effect of digoxigenin-11-dUTP on mol wt and amplification efficiency in PCR labeling. Amplification products were electrophoresed on a 1% agarose minigel and stained with ethidium bromide. The first lane contains molecular markers (1-kbp ladder; Gibco BRL, Gaithersburg, MD). The second lane contains the target plasmid amplified without digoxigenin, and the third lane the plasmid amplified with 10% of the dTTP replaced by dioxigenin-11-dUTP. This high level of dUTP was used to show both the mol-wt shift and decreased amplification efficiency associated with higher levels of probe modification.

5. We have not really found it necessary to purify the labeled probes but cleanup may slightly reduce the background seen with Southern blots. We have found ethanol precipitation to be an effective way to clean up digoxigenin-modified probes.To do this add 3*M* NaOAc to a final concentration of 0.3*M* and 2 vol of ethanol (100%), mix well, and place at −70°C for 15 min. Microcentrifuge for 5 min and discard the supernatant. Wash pellet in 700 μL of 70% ethanol and microcentrifuge for 2 min. Air-dry pellet thoroughly and resuspend in 100 μL TE.

Acknowledgment

This work was supported in part by a USDA Hatch project (ARZT-136440-H-25-042) and by a USDA NRI Competitive Grants Program award (91-37300-6453).

CHAPTER 11

Production of DNA Hybridization Probes with Digoxigenin-Modified Nucleotides by Random Hexanucleotide Priming

Tom McCreery and Tim Helentjaris

1. Introduction

Oligolabeling using random hexanucleotide primers *(1)* is an effective method for producing DNA hybridization probes of high specific activity, although most previous variations have utilized radionuclides as the mechanism for modifying the incorporated nucleotides. With the development of nonradioactive detection methods, the detection is accomplished through the use of either avidin- or antidigoxigenin-alkaline phosphatase conjugates coupled with either a colorimetric or chemiluminescent reaction *(2)*. Both of these strategies require the production of probes with biotin or digoxigenin modifications and demand the crafting of methods for incorporating these modified nucleotides into nucleic acid copies. We have experience with both types of modified probes and have found the digoxigenin-modified probes to be preferable in that they usually cause less nonspecific background signal.

Earlier studies reported that DNA polymerase I could utilize digoxigenin-11-dUTP as a substrate and incorporate it into double-stranded DNA *(3)*. We have similarly found that only simple modifications of our earlier protocols are necessary to permit the production of digoxigenin-modified probes that are capable of detection of very

From: *Methods in Molecular Biology, Vol. 28:*
Protocols for Nucleic Acid Analysis by Nonradioactive Probes
Edited by: P. G. Isaac Copyright ©1994 Humana Press Inc., Totowa, NJ

low amounts of target DNAs (1 pg or less) in a Southern hybridization. In particular this technique is effective for labeling DNA fragments isolated from low-melting-point agarose gels (*see* Note 1) and for labeling DNA fragments that are resistant to PCR amplification. As the modified base is relatively cheap when compared to radionuclide-modified nucleotides and the products are stable for much longer times than ^{32}P-labeled probes, this method offers an economic and safe mechanism for producing hybridization probes that are quite capable of detecting single-copy sequences within very complex genomes.

2. Materials

All oligolabeling reagents are stored at –20°C and thawed on ice immediately prior to use.

1. Solution O: Dissolve 44.52 g of Tris base and 7.47 g of $MgCl_2$ in 150 mL of distilled water, adjust pH to 8.0 with HCl, bring volume up to 250 mL.
2. Solution A: 18 µL ß-Mercaptoethanol, 25 µL of 100 mM dATP, 25 µL of 100 mM dCTP, 25 µL of 100 mM dGTP, 4 µL of 100 mM dTTP, 899 µL of Solution O (*see* Note 2).
3. Solution B (2M HEPES): Dissolve 119.15 g HEPES in 100 mL distilled water, adjust pH to 6.6 with 4M NaOH, and bring volume up to 250 mL.
4. Solution C: Add 555 µL sterile distilled water to 50 U of random hexanucleotides (Pharmacia, Uppsala, Sweden).
5. 5X Oligo Buffer: Mix solutions A, B, and C at a ratio of 1.0:2.5:1.5.
6. BSA: 10 ng/µL.
7. Digoxigenin-11-dUTP: Dilute the 25 µL of 1 mM digoxigenin-11-dUTP stock (Boehringer Mannheim, Mannheim, Germany) to 400 µM by adding 37.5 µL of sterile water.
8. Klenow fragment of DNA polymerase I from *E. coli*: Dilute stock to 1 U/µL with 7 mM Tris-HCl, pH 7.5, 7 mM $MgCl_2$, 50 mM NaCl, 50% glycerol.
9. DNA fragment to be labeled (*see* Note 1).

3. Method

1. Aliquot 50–500 ng of DNA (excised from the vector and isolated by suitable methods on an agarose gel; *see* Note 1) into a 0.65-mL microcentrifuge tube and adjust the volume to 18 µL with water.
2. Heat denature the DNA for 10 min at 95°C.
3. Immediately place the tube containing the DNA on ice.

4. Add the following reagents as quickly as possible and mix thoroughly: 5 µL 5X Oligo buffer, 0.5 µL BSA, 0.5 µL digoxigenin-11-dUTP, 1 µL Klenow.
5. Incubate at 37°C for 90 min to 18 h (3 h seems to provide efficient labeling).
6. The probe may be used immediately or stored either at 4°C (short term) or –20°C (long term) (*see* Notes 3 and 4).

4. Notes

1. Fragments may be isolated from low-melting-point agarose by the following method *(4)*. Pour a low-melting-point agarose gel, load the DNA samples, and run the gel at 4°C to prevent the gel from melting. Stain the gel with ethidium bromide. Use a handheld long wavelength UV source to locate your fragment (this reduces the radiation damage to the DNA). Excise the band with a razor blade or drinking straw. Add 2–3 vol of TE and incubate for 5 min at 65°C to melt the gel. This solution can then be cleaned up further with phenol:chloroform, however, we do not find it necessary.

2. Background may be reduced by decreasing the percentage of modified base in the reaction from 20% of the total dTTP concentration to 5%. This also may increase the efficiency of the reaction as well as utility of the final product. We have found that digoxigenin-modified probes are slightly different chemically when compared to radioactively labeled probes, probably because of the long side chains introduced with the modified base. They form more stable hybrids at lower hybridization temperatures and are more difficult to remove from membranes, once hybridized. Lowering of the dUTP concentration to <10% has a dramatic effect on these qualities.

3. We have not really found it necessary to purify the labeled probes, but cleanup may slightly reduce the background seen with Southern blots. We have found ethanol precipitation to be an effective way to clean up digoxigenin-modified probes. To do this add $3M$ NaOAc to a final concentration of $0.3M$ and 2 vol of ethanol (100%), mix well, and place at –70°C for 15 min. Microcentrifuge for 5 min and discard the supernatant. Wash the pellet in 700 µL of 70% ethanol and microcentrifuge for 2 min. Air-dry the pellet thoroughly and resuspend in 100 µL 10 mM Tris-HCl, pH 8.0, 1 mM EDTA.

4. Unfortunately, there is no quantitative method to measure digoxigenin-incorporation into the final probe, short of spotting the probe onto a membrane and testing with antidigoxigenin/alkaline phosphatase conjugate.

Acknowledgment

This work was supported in part by a USDA Hatch project (ARZT-136440-H-25-042) and by a USDA NRI Competitive Grants Program award (91-37300-6453).

References

1. Feinberg, A. P. and Vogelstein, B. (1983) A technique for radiolabelling DNA restriction fragments to high specific activity. *Analyt. Biochem.* **132,** 6–13.
2. Voyta, J. C., Edwards, B., and Bronstein, I. (1988) Ultrasensitive chemiluminescent detection of alkaline phosphatase activity. *Clin. Chem.* **34,**1157.
3. Kreike, C. M., de Koning, J. R. A., and Krens, F. A. (1990) Non-radioactive detection of single-copy DNA-DNA hybrids. *Plant Mol. Biol. Rep.* **8,**172–179.
4. Sambrook, J., Fritsch, E. F., and Maniatis, T. (1989) *Molecular Cloning: A Laboratory Manual.* Cold Spring Harbor Laboratory, Cold Spring Harbor, NY.

CHAPTER 12

Digoxigenin Labeling of RNA Transcripts from Multi- and Single-Locus DNA Minisatellite Probes

Esther N. Signer

1. Introduction

The technique of DNA fingerprinting detects a large number of independent hypervariable minisatellite loci throughout the genome in a single hybridization experiment and is an extremely powerful and sensitive means for genetic analysis of very many animal and plant species (a broad overview is given in ref. *1*). This technique commonly involves the labeling of DNA probes by incorporation of ^{32}P-tagged nucleotides and autoradiography. However, in some species, such as pigs *(2)* and Galapagos iguanas (K. Rassmann, personal communication and unpublished results), or in cases where only small amounts of DNA on Southern blots are available, greater sensitivity is needed to produce scorable and informative banding patterns. In addition, to make the method suitable for routine screening or processing of large numbers of samples it should be easy to perform, low in health risk, quick, and cheap.

In order to meet the above criteria, RNA transcripts are applied instead of DNA probes and radioactive deoxyribonucleotides are replaced by digoxigenin-labeled ribonucleotides. Transcription by T7 RNA polymerase of a linearized recombinant vector containing the DNA probe and a T7 promoter, in the presence of digoxigenin-

From: *Methods in Molecular Biology, Vol. 28:*
Protocols for Nucleic Acid Analysis by Nonradioactive Probes
Edited by: P. G. Isaac Copyright ©1994 Humana Press Inc., Totowa, NJ

rUTP, produces several mg of highly specific and densely labeled full-length transcripts in 1 h. Further advantages are:

1. The template (and the RNA product before hybridization) does not have to be denatured.
2. The product is easy to purify and to recover.
3. It is stable over several months (at –20°C).
4. In terms of reusing the blots, removal of the probe by sodium hydroxide is rapid and efficient without damage to the DNA.

Compared with DNA, more care has to be taken when working with RNA to prevent any RNase contamination. Use of gloves (hands are the major source of RNase), diethyl pyrocarbonate (DEPC)-treated solutions, disposable tubes, pipet tips from freshly opened bags, and RNasin (a potent ribonuclease inhibitor) in the labeling reaction, minimizes the risk of RNA degradation.

2. Materials

All the enzymes and solutions are stored at –20°C in small aliquots unless where otherwise indicated. Some of the solutions (marked by *) are commercially available (Promega, Madison, WI).

1. Vector/insert template (250 ng/µL) suitable for T7 RNA polymerase: The most widely used multi-locus minisatellites 33.15 and 33.6 for DNA fingerprinting analysis *(3)* have been subcloned into pSPT18 and pSPT19 vectors and are freely available *(4)*. Alternatively, polymorphic DNA markers isolated from genomic libraries and subcloned into the *Eco*RV site of the pBluescript II KS+ vector (Stratagene, La Jolla, CA can also be used as substrates *(5)*.
2. Restriction endonucleases *Eco*RI and *Hind*III (10 U/µL each) and their corresponding 10X incubation buffers (*see* Note 1).
3. Diethyl pyrocarbonate (DEPC). **Caution**: DEPC is a suspect carcinogen and should be used in a fume-hood. Store at 4°C and protect from moisture.
4. DEPC-treated distilled water, 0.05% (v/v). Mix thoroughly in a glass bottle and incubate at room temperature overnight. Autoclave for about 30 min to remove any residual DEPC. Store at 4°C. Prepare or dilute all solutions using this water only.
5. rATP, rCTP, and rGTP, 0.01M each.*
6. 0.01M Digoxigenin-11-rUTP (Boehringer).
7. 5X Transcription buffer*: 0.2M Tris-HCl, 0.03M MgCl$_2$, 0.01M spermidine, 0.05M NaCl, pH 7.5.

8. 0.1*M* Dithiothreitol (DTT).*
9. 40 U/µL RNasin.*
10. 50 U/µL T7 RNA polymerase (*see* Note 2).
11. 0.05*M* EDTA, pH 8.0: This is most conveniently made as a 1:10 dilution from a 0.5*M* stock solution.
12. 7.5*M* Ammonium acetate. Store at room temperature.
13. 100% Ethanol. Store in 50-mL aliquots at –20°C.
14. 80% Ethanol. Store in 50-mL aliquots at –20°C.

3. Method

1. Prepare the template for transcription by linearizing approx 5 µg of vector/insert DNA in a final volume of 20 µL containing 1X incubation buffer and 10 U restriction endonuclease for at least 1 h at 37°C. Use *Hind*III for the vector system pSPT18, *Eco*RI for pSPT19 *(4)*, and *Hind*III for DNA probes cloned into the *Eco*RV site of pBluescript II KS⁺. This will result in the linear order: vector DNA, T7 promoter, insert (=hybridization probe), cleavage site of the endonuclease used. Store the template at –20°C (*see* Notes 3–5).
2. Thaw the following reagents on ice: Linearized DNA template, rATP, rCTP, rGTP, digoxigenin-11-rUTP, 5X transcription buffer, and DTT. Remove enzymes just prior to use, place on ice, and return to the freezer as soon as possible. At room temperature pipet the following amounts into a 1.5-mL screw-cap Eppendorf tube (a snap-cap tube tends to leak during ethanol precipitation): 2 µL linearized DNA template (500 ng), 1.7 µL DEPC-water, 0.5 µL rATP, 0.5 µL rCTP, 0.5 µL rGTP, 1 µL digoxigenin-11-rUTP, 2 µL 5X transcription buffer, 1 µL DTT, 0.5 µL RNasin, and 0.3 µL T7 RNA polymerase. The final reaction volume is 10 µL. Mix briefly by slightly tapping the tube without introducing air bubbles. Give it a short spin and incubate in a 38°C water bath for 1 h (*see* Note 6).
3. Stop the reaction by adding 2.5 µL 0.05*M* EDTA and put the tube on ice.
4. Add 87 µL of DEPC-water and remove unincorporated nucleotides by ethanol precipitation (*see* Note 7). Add 50 µL 7.5*M* ammonium acetate plus 450 µL 100% ethanol (–20°C). Invert the tube a couple of times then leave at –80°C for 30 min.
5. Mark the outside of the tube where the pellet is to be expected then spin in an angled rotor in a bench centrifuge at maximum speed (10,000*g*) for 30–45 min (*see* Note 8).
6. Carefully take off as much supernatant as possible but make sure that the tip docs not touch the pellet.

7. Rinse the pellet with 500 µL ice-cold 80% ethanol, which is added along the tube wall. Do not invert the tube as this might loosen the pellet.

8. Orientate the tube in the same way as for the previous spin and spin for 5 min at full speed. Take off the ethanol as before.

9. Carefully remove any remaining drops by sucking through a sterilized Pasteur pipet connected to a vacuum pump. Do not use a vacuum drier as the pellet might become too dry and resuspension will be very difficult.

10. Resuspend the pellet in 150 µL DEPC-water. Flick the tube several times and leave it on ice for 15 min to dissolve completely (*see* Notes 9 and 10).

11. Either use some aliquots for hybridization (e.g., *see* Chapters 18, 29, and 30) or store the probe at –20°C.

12. In order to verify whether the labeling has worked you may produce a small dot blot from diluted aliquots of the probe and perform color detection as described in Chapter 18.

4. Notes

1. The choice of a particular enzyme depends on the type of vector and the nature of the insert. Use an enzyme that cuts within the multiple cloning site of the vector distal to the insert from the T7 promoter, to produce a linear molecule containing vector DNA, T7 promoter, insert, and cleavage site in the correct transcription direction.

2. The DNA-dependent RNA polymerase from phage SP6 can be used as an alternative to T7 RNA polymerase. Although both enzymes produce RNA probes of equal quality (R. Carter, personal communication), T7 RNA polymerase is preferred as it is less expensive.

3. Standard protocols for rapid extraction of recombinant plasmid DNA from bacterial cultures *(6),* omitting further purification by equilibrium centrifugation in CsCl-ethidium bromide gradients, give DNA substrates of sufficient purity (*see* Chapter 9).

4. The probe/vector DNA template must be linearized to restrict the transcription to the insert. Cotranscription of the vector will reduce the label density of the probe and may cause background.

5. To confirm complete linearization and to estimate the concentration run a small aliquot along with known amounts of a suitable marker (e.g., phage λ DNA cut with *Hind*III) on a 0.8% agarose test gel.

6. The concentrations of unlabeled nucleotides, buffer, and DTT in the reaction mixture, and the incubation conditions are the same as for RNA labeling using ^{32}P rCTP *(4).* Larger amounts of DNA templates can be used by simply scaling up the volumes. Labeled RNA probes produced from 1 µg template (in a reaction volume of 20 µL and finally dissolved

in 300 µL DEPC-water) should be enough for about 20 hybridizations (i.e., forty 20 × 20 cm blots; *see* Chapter 18).

7. The removal of unbound label is essential as in high concentration it hybridizes to the membrane and causes background spots.

8. This speed and time of centrifugation are important to maximize the yield of RNA probe. A small pellet should be visible after centrifugation when held in front of a good light source. Avoid using a swing out rotor as the pellet would be at the very bottom of the tube and therefore difficult to see.

9. The DNA template is copurified with the RNA probe, but it does not interfere with the labeled probe during hybridization. Thus, DNase I digestion of the product is usually not necessary.

10. We do not routinely determine the concentration of the probe, but this can be done either by running an aliquot on a test gel (under RNase-free conditions), which also gives information on the length of the transcript, or by spectrophotometry at 260 nm. This latter method, however, requires DNase I digestion of the product after step 2.

References

1. Burke, T., Dolf, G., Jeffreys, A. J., and Wolff, R. (eds.) (1991) *DNA Fingerprinting: Approaches and Applications.* Birkhäuser Verlag, Basel.
2. Signer, E. N. and Jeffreys, A. J. (1992) Both "hot" and "cold" transcripts of minisatellites 33.15 and 33.6 produce informative DNA fingerprints in pigs. *Fingerprint News* **4/2,** 3–7.
3. Jeffreys, A. J., Wilson, V., and Thein, S. L. (1985) Hypervariable "minisatellite" regions in human DNA. *Nature* **314,** 67–73.
4. Carter, R. E., Wetton, J. H., and Parkin, D. T. (1989) Improved genetic fingerprinting using RNA probes. *Nucleic Acids Res.* **17,** 5867.
5. Signer, E. N. and Jeffreys, A.J. manuscript in preparation.
6. Sambrook, J., Fritsch, E. F., and Maniatis, T. (1989) Large-scale preparations of plasmid DNA, in *Molecular Cloning: A Laboratory Manual,* 2nd ed., vol. 1. Cold Spring Harbor Laboratory, Cold Spring Harbor, NY, pp. 33–39.

CHAPTER 13

Labeling of Double-Stranded DNA Probes with Biotin

Angela Karp

1. Introduction

Probes labeled with biotin have been used for hybridizations to Southern blots and to chromosomes *in situ* since the early 1980s *(1)*. Nowadays, the labeling of double-stranded DNA probes with biotin is made easy by the availability of labeling kits that provide everything required for nick translation of DNA probes in the presence of biotinylated UTP. The method described here is based on the use of the BRL (Bethesda Research Laboratories, Gaithersburg, MD) BIONICK Labeling Kit (catalog no. 8247SA). While the labeling reaction is taking place, a fractionation column is prepared to purify the biotin-labeled DNA. After the fractions have been collected, incorporation is assessed on test strips using a BRL detection kit (catalog no. 8239SA). The whole procedure takes a whole day, with a convenient break when the test strips are incubating in the vacuum oven (*see* step 18 below). Several probes may be labeled in one experiment, although four probes is about the optimum number to handle by a single worker.

2. Materials

1. BRL BIONICK™ Labeling kit (no. 8247SA). This kit contains : 250 μL 10X dNTP mix, 250 μL DNA polymerase I, 250 μL stop buffer, 20 μL control DNA, 1 mL stop buffer.

From: *Methods in Molecular Biology, Vol. 28:*
Protocols for Nucleic Acid Analysis by Nonradioactive Probes
Edited by: P. G. Isaac Copyright ©1994 Humana Press Inc., Totowa, NJ

2. BRL DNA Detection kit (no. 530-8239SA). This kit contains: 120 µL BRL streptavidin, 60 µL biotinylated calf intestinal alkaline phosphatase, 660 µL nitro-blue tetrazolium (NBT), 500 µL 5-bromo-4-chloro-3-indolyl phosphate (BCIP), 25 pg/µL biotinylated DNA.
3. Silanized glassware and Eppendorf microcentrifuge tubes. Fill a glass beaker containing Eppendorfs with "Repelcote" (a silicon treatment water-repellent from BDH) (N.B. this material is harmful and the procedure should be carried out in a fume-hood). Swirl around, discard it, and repeat the treatment. Wash the Eppendorf tubes several times with water before autoclaving and drying in a 80°C oven. Carry out the same procedure for glassware, except after treatment, air-dry, and sterilize in an oven at 180°C.
4. 20X SSC stock solution: 100.2 g trisodium citrate, 175.0 g sodium chloride, 1.0 L double-distilled water, autoclave.
5. 1X SSC + 0.1% SDS: 5 mL 20X SSC, 0.1 g SDS, 95 mL double-distilled water, autoclave.
6. Sephadex G-50: 1 g Sephadex G-50 per 50 mL 1X SSC + 0.1% sodium dodecyl sulfate (SDS).
7. 20 mM Tris, 5 mM EDTA: 20 mL 1M Tris-HCl, pH 7.5, 10 mL 0.5M EDTA, 970 mL double-distilled water, autoclave.
8. Buffer 1: 6.25 mL 1M Tris-HCl, pH 8.0, 6.25 mL 1M NaCl, 0.5 mL 1M MgCl$_2$, 0.125 mL Triton X-100, 237 mL sterile distilled water. Make this just before use (*see* Section 3., step 19).
9. Buffer 2: 0.75 g Bovine serum albumin (BSA) in 25 mL of buffer 1. Make this just before use (*see* Section 3., step 19).
10. Buffer 3: 20 mL 1M Tris-HCl, pH 9.5, 20 mL 1M NaCl, 10 mL 1M MgCl$_2$, 150 mL sterile distilled water. Make this just before use (*see* Section 3., step 19).
11. TE buffer (pH 8.0): 10 mM Tris-HCl, pH 8.0, 1 mM EDTA.
12. Deionized formamide: 200 mL formamide, 1 g Dowex 1 (chloride form) (base anion), 1 g Dowex 50W (hydrogen form). Stir for 1 h. Filter through Whatman filter paper no. 1.
13. NBT and BCIP are supplied in the detection kit but can also be made up as follows: NBT, 75 mg/mL in 70% dimethylformamide; BCIP, 50 mg/mL in dimethylformamide.
14. Bio-Rad (Richmond, CA) Bio Dot Blotting nitrocellulose membrane (or similar) material, e.g., nylon-based hybridization membranes.
15. Dye mix: 7.5 mL Buffer 3, 33 µL NBT, and 25 µL BCIP (NBT and BCIP are in the detection kit). Make this just before use and keep in the dark.
16. 3M Sodium acetate: Add acetic acid to adjust pH to 6.0.
17. Absolute ethanol. Store at –20°C.

3. Method (*see* Note 1)

1. Keeping the contents of the BIONICK kit on ice until they are required, pipet the following into a 1.5-mL silanized Eppendorf (*see* Note 2): 5 μL 10X dNTP mix, 3–4 μL probe DNA (*see* Note 3). Make up the volume to 45 μL by adding autoclaved H_2O from the kit and add 5 μL of the 10X enzyme mix provided.

2. Whirlimix the contents of the Eppendorf tube and centrifuge in a benchtop microfuge at 15,000g for 5 s.

3. Incubate the reaction mix at 16°C for 1 h in a water bath in a cold room.

4. While the labeling reaction is proceeding prepare a fractionation column (steps 5–8) and a test strip (step 9) for each probe being labeled.

5. Take a 1–mL syringe. Plug the bottom with silanized glasswool to a depth of about a 0.1 mL fraction. Do **not** pack the bottom too tightly.

6. Add Sephadex G-50 equilibrated with 1X SSC + 0.1% (w/v) SDS using a Pasteur pipet. Add the first sephadex quickly, tilting the syringe, then top up carefully to avoid air bubbles, which will cause uneven flow. Continue adding small amounts until the column is equilibrated and no more drops come out from the base.

7. Clamp the syringe to a retort stand ready for use.

8. Number six siliconized Eppendorf tubes and place in a collection rack ready for use (*see* Note 4).

9. Now prepare a test strip for each probe. Using the backing paper to avoid directly handling the nitrocellulose, take a sheet of nitrocellulose. Cut a 1-cm strip and, with a sharp pencil, divide the strip into six 1-cm squares. Number each square in the top right hand corner. Label the date and probe code in the first square (note, the first sample is discarded in step 16). Place the test strip(s) on one side ready for use.

10. After the 1 h incubation, add 5 μL of stop buffer from the BIONICK kit to the labeling reaction and mix briefly. Place the Eppendorf tube on ice.

11. Add 15 μL Dextran blue to each Eppendorf tube. Mix well and then pipet all the mixture to the top of the fractionation column.

12. Quickly add 100 μL 1X SSC + 0.1% SDS to the top of the column, washing out the Eppendorf tube first and then adding the remainder to the column. Allow the drops coming from the column to collect into a beaker and discard them.

13. When the drops stop, place the rack of six Eppendorf tubes below the column so that the first one is directly beneath the syringe nozzle.

14. Add 100 μL of 1X SSC + 0.1% SDS to the top of the column and let the drops collect into the first Eppendorf tube. There should be about five to six drops.

15. Move the rack until the second Eppendorf tube is underneath the syringe and add another 100 µL of 1X SSC + 0.1% SDS to the top of the column.
16. Continue until all the blue band has moved off the column and all the Eppendorf tubes have been used (*see* Note 5). Discard the first tube.
17. Lay out the test strip on the bench. Using a clean pipet tip each time, remove a 1-µL aliquot from each collected fraction and spot it into the center of the correspondingly numbered square on the test strip.
18. Bake the filters at 80°C for 1 h in a vacuum oven.
19. Prepare the buffers 1 to 3 for the detection of biotin.
20. After baking the test strip, rehydrate for 1 min in buffer 1 at room temperature.
21. Incubate the strip for 20 min in buffer 2 at 42°C (*see* Note 6).
22. Blot the strip dry between two sheets of filter paper and dry it in the vacuum oven at 80°C for 10–20 min.
23. Rehydrate the strip in buffer 2 for 10 min.
24. Incubate the strip in 2 mL of buffer 1 + 4 µL streptavidin (from the kit) for 10 min with gentle agitation.
25. Wash the strip in 40 mL of buffer 1 for 3 min. Repeat this step twice.
26. Incubate the strip for 10 min with 2 mL buffer 1 + 2 µL kit polyalkaline phosphatase.
27. Wash the strip with 40 mL of buffer 1. Repeat this step.
28. Wash with 40 mL of buffer 3 for 3 min. Repeat this step.
29. Prepare a silver foil-covered small tube.
30. Add the dye mix.
31. Incubate the strip in the dye solution in the foil-wrapped tube for 30 min or until test strip is fully developed (*see* Note 7).
32. Wash the strip with 20 mM Tris, 5 mM EDTA to terminate the reaction.
33. Dry the test strip in the vacuum oven at 80°C for 1–2 min.
34. Based on the results of the test strip (*see* Notes 7 and 8) combine all the labeled fractions you wish to keep into one tube. Usually about four to five fractions are kept (which equals 400–500 µL in total).
35. Add 1/10 vol of 3M sodium acetate (40–50 µL).
36. Add 2 vol of cold absolute alcohol (about 1 mL) and mix by inverting the Eppendorf tube.
37. Freeze the mixture at –70°C for 15 min or at –20°C for 2 h.
38. Centrifuge the sample at 15,000g for 10 min and carefully remove the supernatant.
39. Dry the pellet by placing the tube in a desiccator connected to a vacuum pump for 5 min.
40. Resuspend the pellet in 20 µL of deionized formamide and keep it in the refrigerator at 4°C if it is to be used the next day, or resuspend it in 20 µL of TE buffer and store at –20°C.

4. Notes

1. For safety reasons, wear gloves throughout the whole labeling procedure.
2. It is essential when handling biotin-labeled probes that all glassware and Eppendorf microtubes are silanized.
3. The probe DNA can be in a plasmid vector or used as the excised insert. The concentration should be 1 μg/μL.
4. Standard Eppendorf racks can be used.
5. Usually the bulk of the fraction (visible as a blue band) passes through the column in the third to the fifth Eppendorf tube. If you have packed the silanized wool too tightly, it may take longer to pass through and additional Eppendorf tubes may be required.
6. If you are using nylon membranes and not nitrocellulose, incubate at 65°C.
7. The intensity of color in the spot on the test strip is a good indication of labeling incorporation. Keep only those fractions which gave strong color spots on the strip (usually three to four fractions). Estimations of the levels of incorporation can be achieved by co-incubating a strip with serial dilutions of control biotinylated DNA provided in the kit (*see* kit instructions). However, we have found that provided the color is strong after 30 min to 1 h, there is sufficient incorporation for successful *in situ* hybridization and we do not routinely estimate incorporation levels.
8. If the spots fail to develop any color, check the concentration of your probe DNA and that the NBT and BCIP have not crystallized in the stock bottles (if so, replace with fresh samples).

References

1. Leary, J. J., Brigati, D. J., and Ward, D. C. (1983) Rapid and sensitive colorimetric method for visualizing biotin-labeled DNA probes to DNA or RNA immobilized on nitrocellulose: Bio-blots. *Proc Natl. Acad. Sci. USA* **80,** 4045–4049.

CHAPTER 14

Preparation of Horseradish Peroxidase-Labeled Probes

Ian Durrant and Timothy Stone

1. Introduction

The direct labeling of nucleic acid probes with horseradish peroxidase (HRP) was first described by Renz at EMBL in 1984 *(1)*. The methodology was combined with enhanced chemiluminescence *(2)* (a light producing HRP catalyzed reaction; *see* Chapter 20) allowing the detection of specific hybrids on membranes *(3)*. Further development led to the availability of the first light-based nucleic acid detection system; this was capable of detecting 2.5 pg of target nucleic acid on genomic Southern blots *(4)*. Subsequent protocol and reagent improvements now enable researchers to reliably detect 0.5 pg of target nucleic acid.

The method *(see* Fig. 1) involves the labeling of a single-stranded nucleic acid probe with a positively charged HRP-parabenzoquinone-polyethyleneimine complex (labeling reagent). This complex initially associates with denatured, negatively charged nucleic acid, and is then covalently crosslinked using glutaraldehyde. Probes from 50 bases to 50 kilobases in length have been successfully labeled in this way. The whole labeling protocol involves a few simple steps and takes just 20 min. It is inherently reliable as it is a chemical labeling reaction. The scale of probe labeling can be changed easily to suit the needs of the researcher. Once a probe has been labeled, it can be

From: *Methods in Molecular Biology, Vol. 28:*
Protocols for Nucleic Acid Analysis by Nonradioactive Probes
Edited by: P. G. Isaac Copyright ©1994 Humana Press Inc., Totowa, NJ

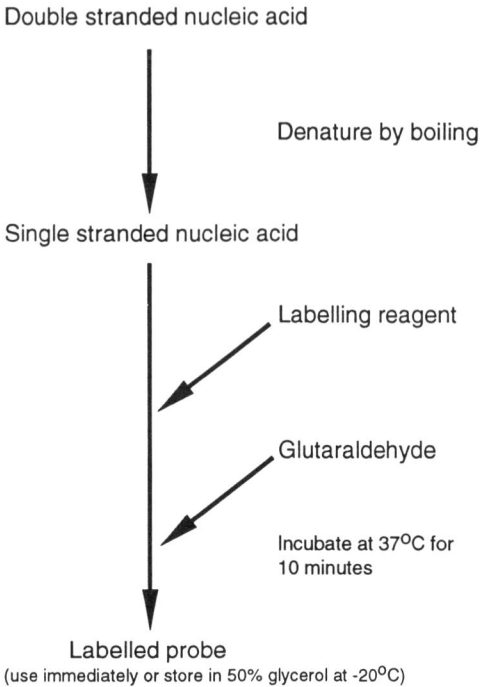

Fig. 1. Outline of the scheme for the production of horseradish peroxidase-labeled probes.

stored for many months. The associated hybridization and post-hybridization protocols are relatively simple, which makes probes labeled directly with HRP particularly suitable for large scale screening, where tens or hundreds of blots are processed weekly.

Probes labeled directly with HRP have been used in many membrane hybridization applications *(5,6)*, including Southern blots, Northern blots, colony and plaque screening, PCR product detection/identification, YAC clone screening, and RFLP analysis.

2. Materials

Reagents 1 and 2 are available in kit form including the reagents for hybridization and chemiluminescent signal generation (ECL™ direct nucleic acid labeling and detection system, RPN 3000, RPN 3001, RPN 3005, Amersham International, Amersham, UK). These materials are stable for up to 3 mo at 4°C. The labeling reagent is light-sensitive and should be protected from intense light or protracted periods of exposure to light.

1. Nucleic acid labeling reagent: charge modified HRP (Amersham).
2. Glutaraldehyde solution: 1.5% (v/v) solution.
3. DNA fragment to be labeled.

3. Method

1. Dilute the DNA to be labeled to a concentration of 10 ng/μL (*see* Note 1) with water (*see* Note 2), in a 1.5-mL conical microcentrifuge tube.
2. Seal the tube and boil for 5 min in a vigorously boiling water bath (*see* Note 3).
3. Cool the DNA on ice for 5 min.
4. Centrifuge the tube briefly (10 s) to settle the liquid at the bottom of the tube.
5. Add an equal volume of DNA labeling reagent. Mix briefly by pipet.
6. Add a volume of glutaraldehyde solution equal to that of the labeling reagent. Mix briefly by pipet.
7. Incubate at 37°C for 10 min.
8. Keep the labeled probe on ice before use (*see* Note 4).

4. Notes

1. A quantity of 10 μL (100 ng) to 150 μL (1.5 μg) of probe can be labeled in each tube. Poor labeling will result if a volume of DNA >150 μL (1.5 μg) is present in each tube.
2. The DNA must be in a low salt buffer (<10 mM NaCl), as ions interfere with the interactions between DNA and labeling reagent thus reducing labeling efficiency.
3. A vigorously boiling waterbath must be used to ensure complete denaturation of double-stranded probes; heating blocks appear not to denature the DNA completely. This denaturation step is not required for single-stranded DNA or RNA.
4. Probes, once labeled, are single stranded and can be used immediately in any membrane hybridization application or may have sterile glycerol added to a final concentration of 50% (v/v) and stored at –20°C for up to 6 mo. Glycerol of the highest quality must be used to avoid inhibition or degradation of the HRP. The recommended probe concentration is 10 ng/mL in hybridizations, although this may be reduced to 2–5 ng/mL for high target applications such as colony and plaque screening (*see* Chapter 20).

References

1. Renz, M. and Kurz, C. (1984) A colorimetric method for DNA hybridization. *Nucleic Acids Res.* **12,** 3435–3444.
2. Whitehead, T. P., Thorpe, G. H. G., Carter, C. J. N., Groucutt, C., and Kricka, L. J. (1983) Enhanced chemiluminescence procedure for sensitive determination of peroxidase-labelled conjugates in immunoassay. *Nature* **305,** 158,159.
3. Pollard-Knight, D., Read, C. A., Downes, M. J., Howard, L. A., Leadbetter, M. R., Pheby, S. A., McNaughton, E., Syms, A., and Brady, M. A. W. (1990) Non-radioactive nucleic acid detection by enhanced chemiluminescence using probes directly labelled with horseradish peroxidase. *Analyt. Biochem.* **185,** 84–89.
4. Durrant, I. (1990) Light based detection of biomolecules. *Nature* **346,** 297,298.
5. Durrant, I., Benge, L. C. A., Sturrock, C., Devenish, A. T., Howe, R., Roe, S., Moore, M., Scozzafava, G., Proudfoot, L. M. F., Richardson, T. C., and McFarthing, K. G. (1990) The application of enhanced chemiluminescence to membrane based nucleic acid detection. *Biotechniques* **8,** 564–570.
6. Stone, T. and Durrant, I. (1991) Enhanced chemiluminescence for the detection of membrane bound nucleic acid sequences, in *Genetic Analysis: Techniques and Applications* **8,** 230–237.

CHAPTER 15

Random Prime Labeling of DNA Probes with Fluorescein-11-dUTP

Bronwen M. Harvey, Claire B. Wheeler, and Martin W. Cunningham

1. Introduction

The selection of an appropriate labeling reagent for a particular experiment, for the most part, depends on the sensitivity and resolution required. For maximum sensitivity in filter hybridizations radioactive labels, such as phosphorous 32, are still the most widely used.

In recent years there has been a trend toward the use of nonradioactive labels because of the disposal and safety advantages. Recent advances in light based detection systems (1) have increased the sensitivity of nonradioactive systems and increased the number of applications for which their sensitivity is adequate.

Nucleic acids can be labeled in several different ways resulting in reporter molecules being incorporated directly into the molecule either uniformly along its length, so maximizing label density, or only at one end. Some reporter molecules, e.g., radioactive or fluorescent tags, can be directly detected. Most others are detected indirectly using enzyme reaction products (2,3). The enzyme, usually horseradish peroxidase or alkaline phosphatase, is often conjugated to a secondary molecule or ligand that ideally has a high affinity for the reporter, e.g., the biotin/streptavidin system (4). Other methods depend on recognition by an enzyme conjugated antibody of a hapten, which is incorporated

From: *Methods in Molecular Biology, Vol. 28:*
Protocols for Nucleic Acid Analysis by Nonradioactive Probes
Edited by: P. G. Isaac Copyright ©1994 Humana Press Inc., Totowa, NJ

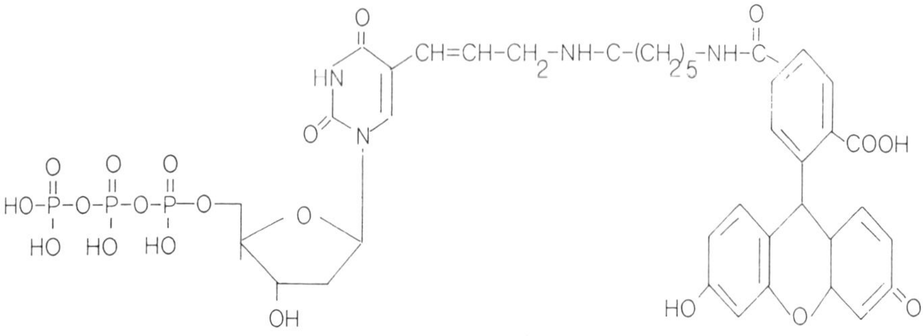

Fig. 1. Fluorescein-11-dUTP.

into the nucleic acid *(5)*. One of the major disadvantages of some nonradioactive labeling systems is that the effectiveness of the labeling reaction cannot be established within a reasonable time before the probe is committed to hybridization. Amersham International (Amersham, UK) has recently introduced a new range of products using hapten-based DNA labeling technology with enhanced chemiluminescence detection *(6)*. A chemically stable hapten nucleotide, Fluorescein-11-dUTP (*see* Fig. 1) can be incorporated into nucleic acid molecules using DNA polymerases, e.g., Klenow polymerase and terminal transferase *(7)*. Subsequently the hapten is detected using a combination of an antifluorescein antibody conjugated to horseradish peroxidase and ECL™. These systems are unique in that they are able to combine the disposal and safety advantages of nonradioactive systems with a rapid semiquantitative assay for incorporation. The assay makes direct use of the physical properties of the fluorescein hapten. After washing away unincorporated fluorescein nucleotide, DNA tagged with fluorescein can be visualized under UV illumination. By reference to a set of standards a semiquantitative estimate of labeling efficiency can be made.

Using the ECL™ random prime system, long probes (minimum length 100 bp) are labeled with fluorescein by the technique of random priming. This procedure was introduced by Feinberg and Vogelstein *(8)* and uses nonamers of random sequences to prime DNA synthesis on a denatured DNA template. The reaction is catalyzed by the Klenow fragment of *E. coli.* DNA polymerase.

2. Materials

Reagents 1–5 are available, fully optimized, from Amersham International plc as ECL random prime labeling and detection system (RPN 3030/3031/3040). The system is extensively tested and necessary controls are also provided. The reagents are stable for at least 3 mo.

1. Nucleotide mix (Amersham): 5X stock solution of dATP, dCTP, dGTP, dTTP, and fluorescein-11-dUTP, in a reaction buffer containing Tris-HCl, pH 7.8, β-mercaptoethanol and magnesium chloride. Store in the dark at –20°C.
2. Primers: random nonamer primers, 50 OD/mL, in an aqueous solution. Store at –20°C.
3. Enzyme solution: 4 U/µL DNA polymerase I Klenow fragment (cloned) in 50 mM potassium phosphate, pH 6.5, 10 mM β-mercaptoethanol and 50% (v/v) glycerol. Store at –20°C.
4. Water: deionized ultrafiltered water.
5. Control fluorescein-labeled DNA, 50 pg/µL. Fluorescein-labeled *Hin*DIII-digested lambda DNA in a solution containing 50 ng/µL herring sperm carrier DNA, 10 mM Tris-HCl, pH 8.0, 1 mM EDTA. Store in the dark at –20°C.
6. 20X SSC: 0.3M trisodium citrate, 3M NaCl, pH 7.0.
7. TE buffer: 10 mM Tris-HCl, pH 8.0, 1 mM EDTA.
8. Positively charged Nylon membrane for example Hybond™-N⁺ (Amersham International plc).
9. 0.5M EDTA, pH 8.0: The pH should be adjusted with NaOH, the EDTA will not fully dissolve unless NaOH is added.
10. Wash buffer: 2X SSC.
11. DNA fragment to be labeled.

3. Method

3.1. Labeling

1. Dilute the template DNA to be labeled to a concentration of 2–25 ng/µL in either water or TE buffer (*see* Notes 1 and 2).
2. Place the required tubes of labeling components, with the exception of the enzyme, in an ice bath to thaw. The enzyme should be left at –20°C until required, and returned to the freezer immediately after use.
3. Denature the template DNA by heating for 5 min in a boiling water bath, then chill immediately on ice. It is strongly advised to denature at least 20 µL.

4. To a 1.5-mL microcentrifuge tube, placed in an ice bath, add the appropriate volume of each reagent in the following order: water to ensure a final reaction volume of 50 µL, 10 µL nucleotide mix, 5 µL of primers, 50 ng of denatured DNA (minimum), 1 µL (4 U) of enzyme (Klenow fragment) (*see* Note 3).

5. Mix gently by pipeting up and down, and cap the tube. Spin briefly in a microcentrifuge to collect the contents at the bottom of the tube (*see* Note 4).

6. Incubate the reaction mix at 37°C for 1 h or at room temperature overnight (*see* Note 5). Stop the reaction by adding 2 µL 0.5*M* EDTA.

7. The labeled probe is now ready for use in hybridization (*see* Note 6). Further purification of the labeled probe is not required, saving loss of valuable probe and time. Probes can be stored in the dark at –20°C for at least 3 mo, avoiding frequent labeling reactions.

8. If desired the reaction can be checked using the rapid labeling assay, or an estimate of probe concentration can be made by preparing serial dot blots, and detecting alongside the control fluorescein-labeled DNA (the direct dot assay).

3.2. Monitoring Incorporation Using a Rapid Labeling Assay

1. Cut a sheet of nylon membrane allowing 1 cm between each sample. Use a pencil to indicate the sample positions.

2. Prepare a negative control consisting of a 1 in 5 dilution, in TE buffer, of the nucleotide mix.

3. With the nylon membrane on a clean nonabsorbent surface (such as SaranWrap™), carefully apply 5 µL of each labeling reaction in the appropriate position. In addition apply 5 µL of the negative control (*see* Note 7).

4. Allow the samples to absorb for 1 min and then immerse the membrane in 100 mL of prewarmed wash buffer and incubate at 60°C with gentle shaking for 10–15 min (*see* Note 8).

5. Remove the membrane with clean forceps and lay it on a sheet of absorbent filter paper.

6. Visualize the membrane DNA side down on a UV transilluminator (*see* Note 9). The fluorescein-labeled DNA can be seen as a yellow/green fluorescent ring. Little or no fluorescein should be observed on the negative control. A semiquantitative estimate of labeling efficiency can also be obtained (*see* Note 10).

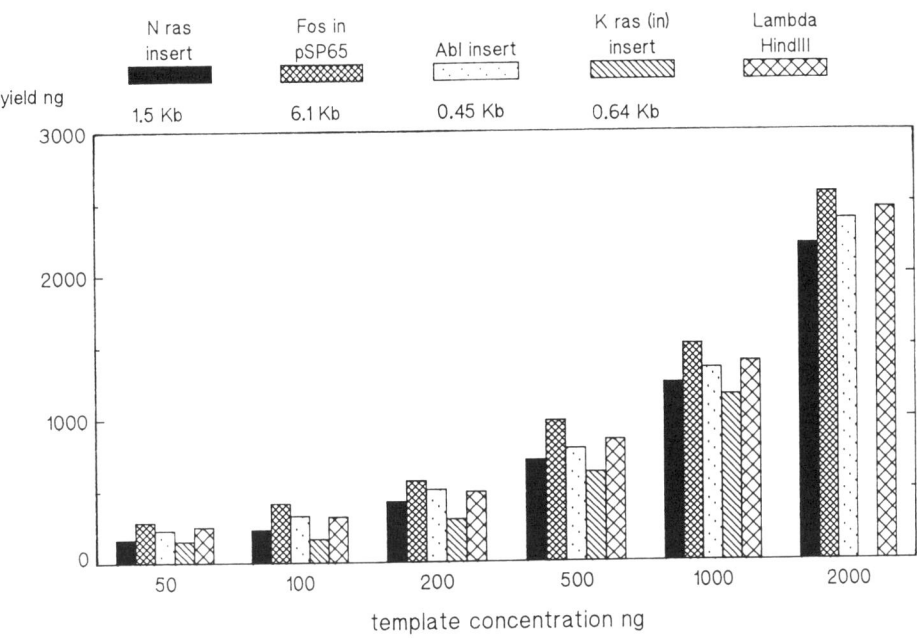

Fig. 2. Effect of template type and concentration on yield from an ECL random prime reaction. Yields given are total amounts of DNA present (labeled DNA plus template) following each labeling reaction. The reactions were carried out for 1 h at 37°C. The N-ras, abl, and K-ras proto-oncogene inserts were excised from pSP65 plasmid by *Eco*RI, separated from the plasmid by gel electrophoresis, and purified using DEAE-cellulose paper.

3.3. Monitoring Incorporation Using a Direct Dot Assay

1. Prepare serial dilutions of the labeled probe and the control fluorescein-labeled DNA down to a concentration of 0.1 pg/μL.
2. Apply 1-μL aliquots of each dilution to a sheet of nylon membrane; allow the dots to air-dry. The DNA can then be fixed by UV or baking for 2 h at 80°C.
3. The fluorescein is detected, after membrane blocking, using the antifluorescein/horseradish peroxidase conjugate and ECL detection. Full details of the detection procedure can be found in Chapter 21.

4. Notes

1. A wide range of DNA concentrations can be used in the labeling reaction (*see* Fig. 2). There can be considerable net synthesis of probe in

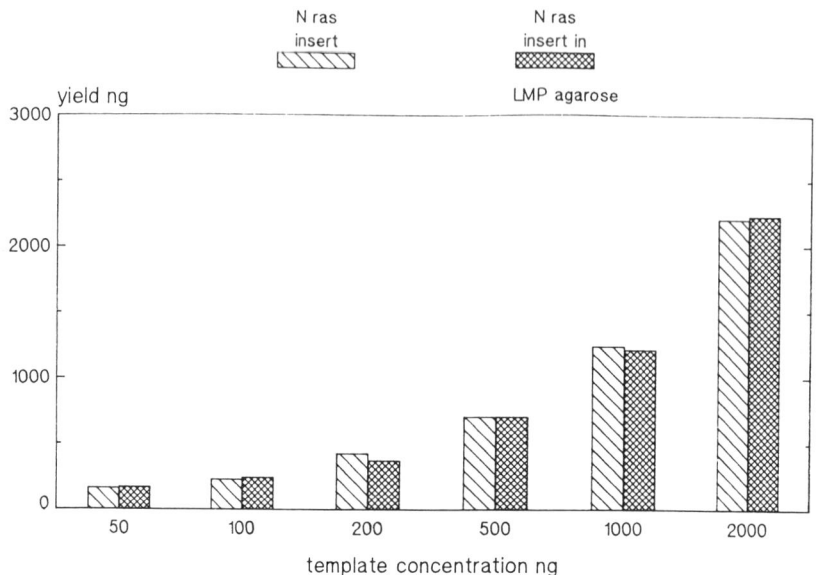

Fig. 3. Effect of template purity on the yield from an ECL random prime reaction. Total amounts of DNA present (labeled DNA plus template) following labeling reactions for 1 h at 37°C with various quantities of template. 5 µL of N-ras proto-oncogene insert prepared in low-melting-point agarose were used in the test reactions.

contrast to a radioactive random prime reaction, since none of the nucleotides is present at a limiting concentration. With Klenow polymerase the newly synthesized DNA strand can be progressively displaced by an adjacent growing strand so that more than a single copy of the initial template can be made. This can result in a several fold increase in the amount of DNA present as the reaction progresses. Synthesis is most efficient for lower levels of template, although net synthesis tends to increase with the amount of template.

2. A wide variety of DNA can be labeled using the ECL random prime system. Closed circular double stranded DNA can be used as a template. The reaction yield will, however, be improved if the DNA is first linearized since renaturation will be inhibited. Single-stranded DNA, miniprep DNA *(9)*, and inserts in low-melting-point agarose (*see* Fig. 3) can also be effectively labeled, although in some cases a longer labeling reaction may be necessary.

3. The size of the probe is determined by the ratio of the primer to template and by the nucleotide concentration. These have been fully opti-

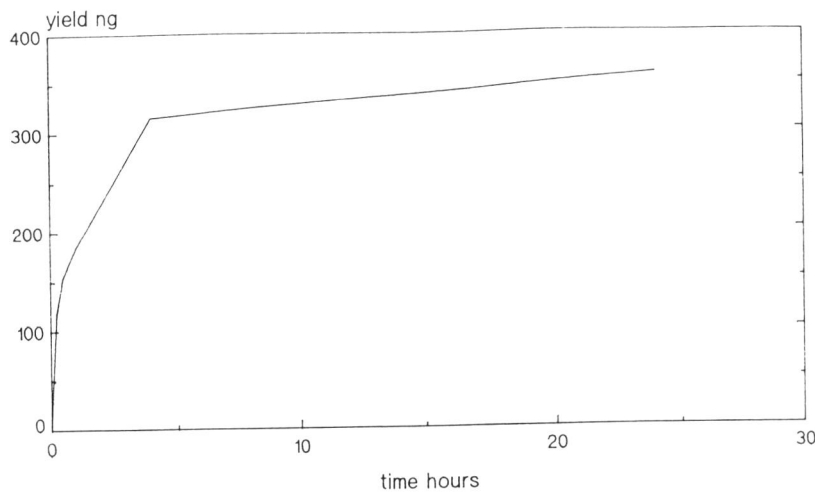

Fig. 4. Kinetics of random primed labeling with fluorescein-11-dUTP. The yields given are total amount of DNA present (labeled DNA plus template) following each standard labeling reaction. Incubation at room temperature.

mized in the ECL random prime labeling system to generate probes with sizes (100–300 bp) suitable for filter hybridizations.

4. Vigorous mixing must be avoided; the resulting shearing forces will result in a loss of enzyme activity.

5. Temperature of incubation and reaction time can be chosen for convenience as the reaction reaches a plateau and does not decline significantly overnight at room temperature (*see* Fig. 4). However the rate of reaction does depend to some extent on DNA purity. For example, with miniprep DNA or DNA in agarose, a longer incubation period may be required. With purified DNA a slight increase in probe yield can be obtained with longer reaction times.

6. In general the final concentration of probe in the reaction mix after labeling is between 3 and 6 ng/μL when starting with 50 ng of template. Reaction yields generally reflect the size of the template; larger templates will tend to produce higher yields.

7. A more discrete ring can be obtained by applying the sample in two 2.5-μL aliquots leaving 1–2 min between additions.

8. As the DNA has not been fixed to the membrane it is necessary to avoid the use of SDS in the wash buffer as this will strip off the labeled probe as well as the free nucleotide. More rapid removal of the unincorporated nucleotide can be obtained by use of a higher salt concentration, for example 5X SSC.

9. The absorbence maximum for fluorescein is 493 nm, so the efficiency of this method may be lower with short wavelength UV transilluminators.
10. To obtain a semiquantitative estimate of labeling efficiency the fluorescein can be compared to that of a dilution series of the nucleotide mix. Prepare 1/25, 1/50, 1/100, 1/250, 1/500, and 1/1000 dilutions in TE buffer, and apply 5 µL of each onto a sheet of nylon membrane (as described in Section 3.2.). Visualize, without washing, on the transilluminator alongside the prepared reaction samples. The labeling reaction has worked acceptably if the intensity is between that of the 1/50 and the 1/250 dilution. It may, however, be possible to successfully use fluorescein-labeled probes in filter hybridizations that lie outside this range. The reference filter can be stored for a week at –20°C in the dark.
11. Correspondence concerning this protocol should be addressed to B. M. Harvey.

References

1. Whitehead, T. P., Thorpe, G. H. G., Carter, C. J. N, Groucutt, C., and Kricka, L. J. (1983) Enhanced luminescence procedure for sensitive determination of peroxidase-labelled conjugates in immunoassay. *Nature* **305,** 158,159.
2. Renz, M. and Kurz, C. (1984) A colorimetric method for DNA hybridization. *Nucleic Acids Res.* **12,** 3435–3444.
3. Pollard-Knight, D., Read, C. A., Downes, M. J., Howard, L. A., Leadbetter, M. R., Pheby, S. A., McNaughton, E., Syms, A., and Brady, M. A. W. (1990) Nonradioactive nucleic acid detection by enhanced chemiluminescence using probes directly labeled with horseradish peroxidase. *Anal. Biochem.* **185,** 84–89.
4. Langer, P. R., Waldrop, A. A., and Ward D. C. (1981) Enzymatic synthesis of biotin labelled polynucleotides: Novel nucleic acid affinity probes. *Proc. Natl. Acad. Sci. USA* **78,** 6633–6637.
5. Simmonds, A. C., Cunningham, M., Durrant, I., Fowler, S. J., and Evans, M. R. (1991) Enhanced chemiluminescence in filter based DNA detection. *Clin. Chem.* **37,** 1527,1528.
6. Cunningham, M. W. (1991) Nucleic acid detection with light. *Life Sci.* **6,** 2–5.
7. Bollum, F. J. (1974) Terminal deoxynucleotidyl transferase, in *The Enzymes,* vol. 10 (Boyer, P. D., ed.), Academic, New York, pp. 145–171.
8. Feinberg, A. P. and Vogelstein, B. (1983) A technique for radiolabelling DNA restriction endonuclease fragments to high specfic activity. *Anal. Biochem.* **132,** 6–13.
9. Holmes, D. S. and Quigley, M. (1981) A rapid boiling method for the preparation of bacterial plasmids. *Anal. Biochem.* **144,** 193–197.

CHAPTER 16

Labeling of Oligonucleotides with Fluorescein

Patricia M. E. Chadwick and Ian Durrant

1. Introduction

Oligonucleotides are a convenient alternative to conventional, cloned DNA probes since they can be rapidly synthesized in large quantities relatively cheaply. They can be made to very specific regions of the genome, without reference to restriction enzyme sites, and are therefore of particular value in the analysis of gene families where sequences may be highly conserved. Oligonucleotides are stable, do not self-anneal, and generally cause fewer problems with nonspecific binding. They are amenable to several labeling techniques, both radioactive and nonradioactive; such methods include chemical modification during synthesis and the enzymatic incorporation of labeled nucleotides at the 5' end or the 3' end.

Terminal transferase is an enzyme that catalyzes the addition of deoxynucleotide triphosphates to the 3' hydroxyl end of double- or single-stranded DNA *(1)*; it can therefore be used to add tails of labeled nucleotides to DNA molecules. This chapter describes a method for the introduction of a tail of fluorescein-dUTP onto the 3' end of an oligonucleotide.

The optimized reaction protocol yields a tail length that is sufficient to ensure high sensitivity in hybridizations, without compromising the stringency of the specific probe sequence *(2)*. The labeling procedure is simple,

From: *Methods in Molecular Biology, Vol. 28:*
Protocols for Nucleic Acid Analysis by Nonradioactive Probes
Edited by: P. G. Isaac Copyright ©1994 Humana Press Inc., Totowa, NJ

reliable, versatile in scale, and rapid, typically taking only 1 h to label 25–250×10^{-12} mol of probe. Purification of the labeled probe to remove unincorporated fluorescein-dUTP is not necessary, and probes may be stored at $-20°C$ in the dark for at least 6 mo.

2. Materials
2.1. Labeling Reaction

Components 1–5 should be stored at $-20°C$. Solutions 6–8 are stable at room temperature. Components 1–5 are available from Amersham International (Amersham, UK) as the ECL™ 3'-Oligolabeling System, RPN 2130.

1. Fluorescein-11-dUTP. Concentrated stock of labeled nucleotide in 10 mM Tris-HCl, pH 8, 1 mM EDTA. Store in the dark (*see* Note 1).
2. Cacodylate buffer: 1.4M sodium cacodylate, pH 7.2, 10 mM cobalt chloride, 1 mM dithiothreitol. (**Warning:** Cacodylate is highly toxic by contact with skin or if swallowed. It may be a carcinogen and with danger of cumulative effects.)
3. Terminal transferase enzyme, 2 U/µL, pH 7.0.
4. Water: sterile, distilled.
5. Oligonucleotide to be labeled: 100×10^{-12} mol are required (*see* Notes 2 and 3).
6. TE buffer: 10 mM Tris HCl, pH 8.0, 1 mM EDTA
7. 20X SSC: 3M NaCl, 0.3M trisodium citrate, pH 7.0.
8. 10%(w/v) SDS.

2.2. Rapid Assay of Probe Labeling

1. Whatman DE81 paper.
2. Wash solution: 2X SSC, 0.1% (w/v) SDS.

3. Method
3.1. Labeling Reaction

1. Place all the components, except the terminal transferase enzyme, on ice to thaw. Leave the enzyme at $-20°C$ until required and return to the freezer immediately after use.
2. To a 1.5-mL conical microcentrifuge tube, on ice, add the labeling reaction components in the following order (*see* Notes 2 and 3): x µL oligonucleotide (100×10^{-12} mol), 10 µL fluorescein-11-dUTP, 16 µL cacodylate buffer, y µL water, 16 µL terminal transferase. The volumes represented by x and y, above, should be selected such that the final volume is 160 µL.

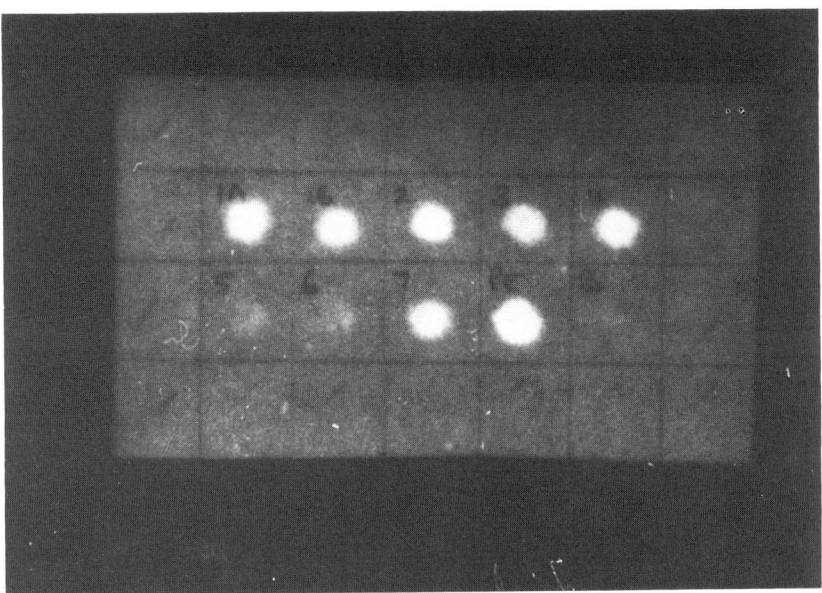

Fig. 1. Rapid labeling assay. Probes were labeled according to the standard protocol and analyzed by the rapid labeling assay described. Positions marked 1A, 1B, 2, 3, 4, 7, and PE represent successful labeling reactions; positions marked 5 and 6 show labeling reactions that were produced by a suboptimal reaction procedure; position marked B is a negative control produced with labeled nucleotide only. The relative intensity of signal results is representative of the sensitivity achieved with these probes in hybridizations.

3. Mix gently with a pipet tip.
4. Incubate the reaction mixture at 37°C for 60–90 min, preferably in the dark.
5. Place the labeled probe on ice for immediate use, or at –20°C for longer term storage.
6. The efficiency of the labeling may be checked by use of a rapid assay that exploits the fluorescent properties of the hapten *(3)*.

3.2. Rapid Assay of Probe Labeling

1. Cut a piece of Whatman DE81 chromatography paper large enough for the number of samples to be checked, allowing 1 cm between samples.
2. Prepare a 1 in 16 dilution of fluorescein-11-dUTP in TE buffer as a blank.

3. Apply 5 µL of each labeling reaction and the blank to the paper. As soon as the samples are absorbed (1 min) immerse the paper in 100 mL of prewarmed wash solution and incubate at 60°C with gentle agitation for 15 min. Do not allow the paper to dry before immersion as unincorporated nucleotides will become difficult to remove.

4. Pour off the wash solution, rinse the paper in water briefly, and replace with ethanol. Remove the paper and blot excess liquid on absorbent paper.

5. View on a 302-nm UV transilluminator: Fluorescein-labeled DNA should appear as yellow/green fluorescent dots. Little or no fluorescence should be visible on the blank sample (*see* Note 4). The paper can be photographed using a Kodak Wratten No. 9 yellow filter on, e.g., Polaroid 667 black and white film (*see* Fig. 1). A semiquantitative estimate of labeling efficiency can be obtained (*see* Note 5).

4. Notes

1. The fluorescein hapten is sensitive to prolonged exposure to light. It is recommended that, for long term storage, solutions containing fluorescein are placed in the dark. However, no special precautions are necessary during the labeling procedure.

2. It is recommended that the oligonucleotide is dissolved or diluted in sterile distilled water.

3. The weight corresponding to 100×10^{-12} mol of an oligonucleotide is dependent on the length of the sequence. Multiplying the number of bases in the oligonucleotide by 33.3 gives the number of nanograms equivalent to 100×10^{-12} mol; e.g., 660 ng of a 20-mer is required for the standard protocol. Although the protocol is written for the labeling of 100×10^{-12} mol of probe, alternative quantities of probe may be prepared by scaling the components up or down in proportion to the amount of oligonucleotide. It is not recommended to exceed 500×10^{-12} mol in a single reaction.

4. If significant fluorescence remains on the blank, and the paper has not dried out, rewash it in 2X SSC, 0.1% SDS at 60°C for a further 15 min, and repeat Section 3.2., steps 4 and 5.

5. To obtain a semiquantitative estimate of labeling efficiency, compare the fluorescence to that of a dilution series of fluorescein-11-dUTP. Prepare 1/100, 1/250, 1/500, 1/1000, and 1/2000 dilutions of the nucleotide in TE buffer and apply 5 µL of each dilution to a piece of DE81 paper. View and photograph directly on a transilluminator, without washing. The labeling reaction has worked acceptably if the intensity of the probe is between that of the 1/500 and 1/2000 dilutions.

References

1. Bollum, F. J. (1974) Terminal deoxynucleotidyl transferase, in *The Enzymes,* vol. 10 (Boyer, P. D., ed.), Academic, New York, pp. 145–171.
2. Durrant, I. (1991) Stringency control with 3' tailed oligonucleotide probes. *Highlights* **2,** 6,7.
3. Cunningham, M. W. (1991) Nucleic acid detection with light. *Life Sci.* **6,** 1–5.

CHAPTER 17

Hybridization of Digoxigenin-Labeled Probes to Southern Blots and Detection by Chemiluminescence

Tom McCreery and Tim Helentjaris

1. Introduction

Probes prepared with either digoxigenin- or biotin-modified nucleotides can be hybridized to Southern blots to detect target nucleic acid sequences. These methods offer an attractive alternative to "radioactively tagged" probes in terms of safety, cost, and efficiency. Most previous nonradioactive strategies utilized the detection of the modified base by the use of a coupled antibody- or avidin-alkaline phosphatase. The blot with the bound alkaline phosphatase was then treated with a compound, such as Nitro Blue Tetrazolium (NBT), that was converted to an insoluble, colored compound at the site of hybridization, thus facilitating visualization of the hybridized probe.

Replacement of this colorimetric reaction with a chemiluminescent process provides for better sensitivity as well as easier reusability of membranes. The development of compounds, such as adamantyl 1,2 dioxetane phosphate (AMPPD) *(1)* provides an easy alternative to previous detection schemes as this compound is also destabilized by alkaline phosphatase, resulting in the production of light, which will expose a standard X-ray film. Probes produced by exactly the same methods as for NBT detection are perfectly usable in this process. The membranes with the bound alkaline phosphatase are soaked in a

From: *Methods in Molecular Biology, Vol. 28:*
Protocols for Nucleic Acid Analysis by Nonradioactive Probes
Edited by: P. G. Isaac Copyright ©1994 Humana Press Inc., Totowa, NJ

dilute solution of AMPPD and then exposed to film at room temperature or 37°C instead of –70°C. Exposure times of 45 min to several hours are then all that is necessary to detect single-copy sequences even in genomic DNA preparations from organisms with very large genomes.

2. Materials

1. Membrane of interest (*see* Chapter 4): These should be prepared as they would be for radioactive detection; however, it is very important to handle the gels and membranes only with gloved hands or forceps as fingerprints cannot be removed and will be made very obvious by the detection protocol. Only nylon membranes should be used as the AMPPD is stabilized in nylon and will not provide the same level of sensitivity when nitrocellulose or polyvinylidene fluoride (PVDF) membranes are used. In addition to the DNA samples, prelabeled mol-wt standards can also be run on the gel and transferred to the membrane (*see* Note 1).
2. Prehybridization/hybridization buffer: 5X SSC, 0.1% *n*-lauroylsarcosine (powdered form), 0.02% sodium lauryl sulfate (SDS), 0.1% blocking agent (either from Tropix or Boehringer Mannheim, Mannheim, Germany), 100 µg/mL salmon sperm DNA (optional, may reduce nonspecific binding).
3. Washing solution: 0.1X SSC, 0.1% SDS.
4. Buffer 1: 0.05M Tris-HCl, pH 7.5, 0.15M NaCl, a 10X solution of this buffer can be made up, autoclaved, and diluted immediately before use.
5. Blocking buffer (Buffer 2): Buffer 1 with 0.1% blocking agent added. It is best to prepare enough buffer 1 to make the blocking buffer the previous day and put it in a 65°C incubator overnight. This allows you to make the blocking buffer just before you use it (*see* Note 2).
6. Antidigoxigenin solution: This solution consists of 50 mL of blocking buffer with 2.5 µL of antidigoxigenin/alkaline phosphatase conjugate (Boehringer Mannheim [Mannheim, Germany] #1093 274). This solution should be made up immediately before use.
7. 1M MgCl$_2$, autoclaved.
8. 5M NaCl, autoclaved.
9. Buffer 3: 2.4 mL diethanolamine, 2.5 mL 1M MgCl$_2$, 2.5 mL 2% Na-azide, 42.6 mL distilled water, pH 10 (*see* Note 3).
10. AMPPD solution: Add 5 µL of Tropix AMPPD to each mL of buffer 3. 0.4 mL of this solution is sufficient for a 100 cm^2 blot.
11. TE: 10 mM Tris-HCl, pH 8.0, 1.0 mM EDTA. Filter and autoclave.

3. Method

1. Prehybridize the blots at 60–65°C in either plastic bags or plastic boxes for 30 min using 3 mL of prehybridization solution/100 cm^2.
2. Denature the probes by heating to 95°C for 10 min. Add the probes to the hybridization solution at a concentration of 5–20 ng/mL of hybridization solution. Generally 2 µL of probe/mL of hybridization solution will produce good results.
3. Replace the prehybridization solution with fresh hybridization solution containing the denatured probe. Incubate from 6 h to overnight at 60–65°C. Mixing during hybridization is not necessary.
4. Remove the hybridization solution (this can be stored at 4°C and re-used several times).
5. Wash the membranes twice for 3 min with at least 200 mL of washing solution with good agitation at room temperature.
6. Wash the membranes for 30 min at 60°C with at least 200 mL of washing solution and with good agitation. This step removes any probe that is not bound to the membrane.
7. Wash the membranes with buffer 1 for 5 min at room temperature (this and all subsequent washes are performed at room temperature and with good agitation). This step removes the washing solution from the membranes and prepares the surface for the blocking agent.
8. Wash the blots with at least 200 mL of blocking buffer for 30 min. This blocks the surface of the membrane to prevent nonspecific binding.
9. Incubate the blots in approx 200 mL of antidigoxigenin solution for 30 min. At this point in the process the antibody/alkaline phosphatase conjugate binds to the digoxigenin-labeled probe that is bound to the membrane.
10. Wash the blots three times for 5 min each wash with 200 mL of buffer 1. These washes remove the unbound antibody/alkaline phosphatase conjugate.
11. Wash blots once for 10 min with buffer 3. This wash ensures that the pH of the blot is appropriate for exposure to AMPPD.
12. Spot 0.4 mL of the AMPPD solution per 100 cm^2 of blot. The alkaline phosphatase portion of the conjugate will now begin to break down the AMPPD and produce light.
13. Wrap the blots in plastic wrap (we have used both Polyvinyl chloride and polyethylene films), smooth to distribute AMPPD solution, and place them in a cassette (do not use an intensifying screen) with standard X-ray film.
14. Expose the film for either 3 h at 37°C or overnight at room temperature. The exposure time depends on the level of target sequence, the percentage of modified base in the probe, and the amount of probe added.

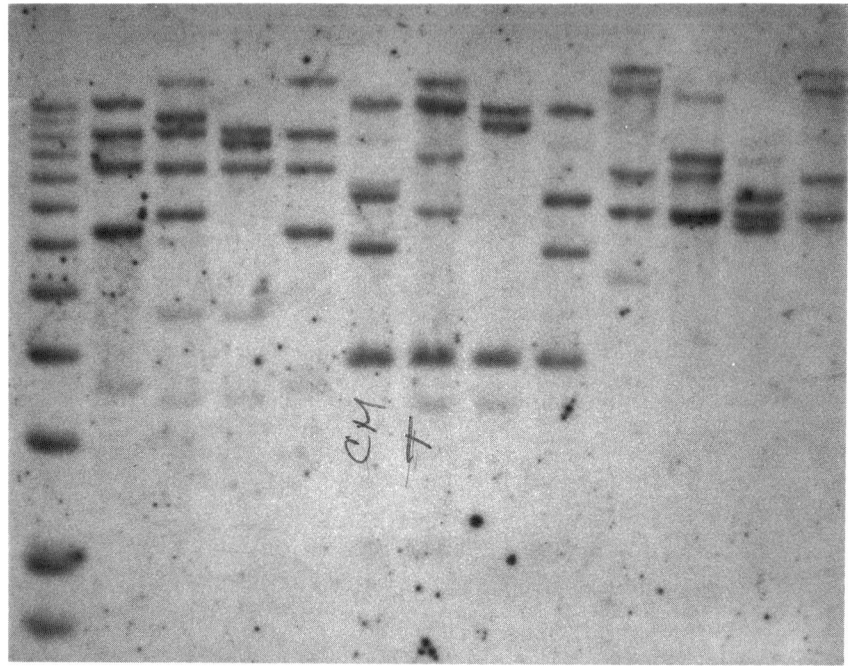

Fig. 1. Southern hybridization with a digoxigenin labeled probe. This is a luminogram of a unique sequence probe labeled with 5% digoxigenin modified base and hybridized to a maize genomic Southern as described. With a 3-h exposure at 37°C single copy fragments are readily detectable.

15. Develop the film. The result should be sharp black bands on a clear background (*see* Fig. 1 and Notes 4 and 5).
16. To reuse the blots, remove the probe by boiling for 10 min in washing solution. Rinse the blots in TE and allow them to air-dry (*see* Note 6).

4. Notes

1. One difference with this type of system is that it is convenient to prepare and utilize a digoxigenin-labeled mol-wt marker in the agarose gel. We find that use of the Klenow fragment of DNA polymerase I to "fill in" the ends of DNA fragments created by restriction digestion is an efficient method for creating mol-wt standards that will be visible every time the blot is treated with alkaline phosphatase/antidigoxigenin conjugate. We particularly like the 1-kb mol-wt ladder from BRL, which already possesses accessible ends and yields a set of mol-wt standards every 1000 bp.

2. If the blocking agent is not fully dissolved, then the final exposed film may show a speckled appearence (*see* Note 5c)
3. If buffer 3 is not made up correctly, or the AMPPD is breaking down, the final film may show a gray background (*see* Note 5b)
4. If there is no signal or a very weak signal:
   a. The probe was not properly prepared or added at too low a level. Check probe on gel to determine whether labeling worked and what the proper concentration of probe should be.
   b. The DNA on blot is at too low a concentration or is degraded. This is a fairly common problem; try adding more DNA per lane. Checking the quality of the DNA you are using by running it undigested on a low percentage agarose gel (*see* Chapters 2 and 3).
   c. AMPPD was too old or degraded. If the labeled markers show up properly this is not the problem and you should look elsewhere.
   d. Antibody was not added or was bad. If the markers show up, this is not the issue and you need to look elsewhere; if you do not see the markers, then perhaps you should check both your conjugate and AMPPD.
5. If there is high background (this is the more common problem with this technique):
   a. Smudges are visible: Check to see if they are fingerprints; you can never remove these from the blot.
   b. The blot has an even gray background or is very dark: This is usually caused by the AMPPD breaking down. This will happen if buffer 3 is improperly made up. If the AMPPD stock solution is bad, you will start to see an even background coming up over the entire blot and signal decreasing; try making this up again.
   c. Spots: These have a number of possible sources and are often noted in the absence of added probe indicating problems with the Ab binding nonspecifically to the blots; we have seen that using "dirty" holders for the blots during film exposure has caused the AMPPD to autocatalytically break down, presumably owing to exogenous alkaline phosphatase or other contaminants; alternatively this is often associated with the buffer 2 if the blocking agent is either not completely in solution or the blocking buffer was not made up immediately before use.
   d. Lane background is high: When the rest of the blot seems relatively clean but there is high lane background overshadowing your bands, this is a characteristic of the probe used and does not reflect upon the conditions; if you want to decrease this problem, the only practical alternative is to use higher stringency washing conditions (65–68°C and 0.1X SSC)
   e. Scratches are visible: When processing the blots, you must be careful not to scratch the surface with forceps; this is certainly not a

problem with radioactive probes but any rough handling of these membranes can cause higher background.

6. If ghost bands are left over from previous uses, it may be improper stripping. Try stripping again. If that does not work then it is probably caused by allowing the blots to dry with probes hybridized to them: Do not allow the blots to dry or just sit too long while probes are still hybridized to them; they cannot be removed and will be on for nearly the life of the blot. Be more careful to seal them during film exposure and strip immediately after taking the blots off film.

Acknowledgment

This work was supported in part by a USDA Hatch project (ARZT-136440-H-25-042) and by a USDA NRI Competitive Grants Program award (91-37300-6453). We are indebted to D. Hoisington for many suggestions that were incorporated into these protocols.

References

1. Voyta, J. C., Edwards, B., and Bronstein, I. (1988) Ultrasensitive chemiluminescent detection of alkaline phosphatase activity. *Clin. Chem.* **34,** 1157.

CHAPTER 18

Southern Blot Hybridization of Digoxigenin-Labeled RNA Minisatellite Probes and Color Detection

Esther N. Signer

1. Introduction

The following procedure describes the hybridization of digoxigenin-labeled RNA minisatellite probes (*see* Chapter 12) to Southern blots of different species and visualization of the DNA banding patterns by color detection. The method can be applied to both multi- and single-locus probes (*see* Figs. 1 and 2, respectively) and also for blots that have been screened several times with both radioactively labeled or digoxigenin-labeled probes *(1)*.

In contrast to radioactively labeled DNA probes, the medium for prehybridization and hybridization is adapted to RNA *(2)* mainly to prevent contamination by RNases. However, diethyl pyrocarbonate (DEPC) treatment of the water is not necessary. Good quality distilled water can be used instead to prepare the solutions as some contain SDS, which inhibits RNase activity (washing solutions) whereas others are kept as highly concentrated stock solutions (20X SSC), are autoclaved and stored at 4°C (TN buffer, alkaline TNM buffer, and TE buffer), or are made freshly (SSC dilutions, blocking buffer, ABC solution, and color solution).

As background artifacts can be a problem in nonradioactive detection systems, any damage of the membrane surface should be avoided

From: *Methods in Molecular Biology, Vol. 28:*
Protocols for Nucleic Acid Analysis by Nonradioactive Probes
Edited by: P. G. Isaac   Copyright ©1994   Humana Press Inc., Totowa, NJ

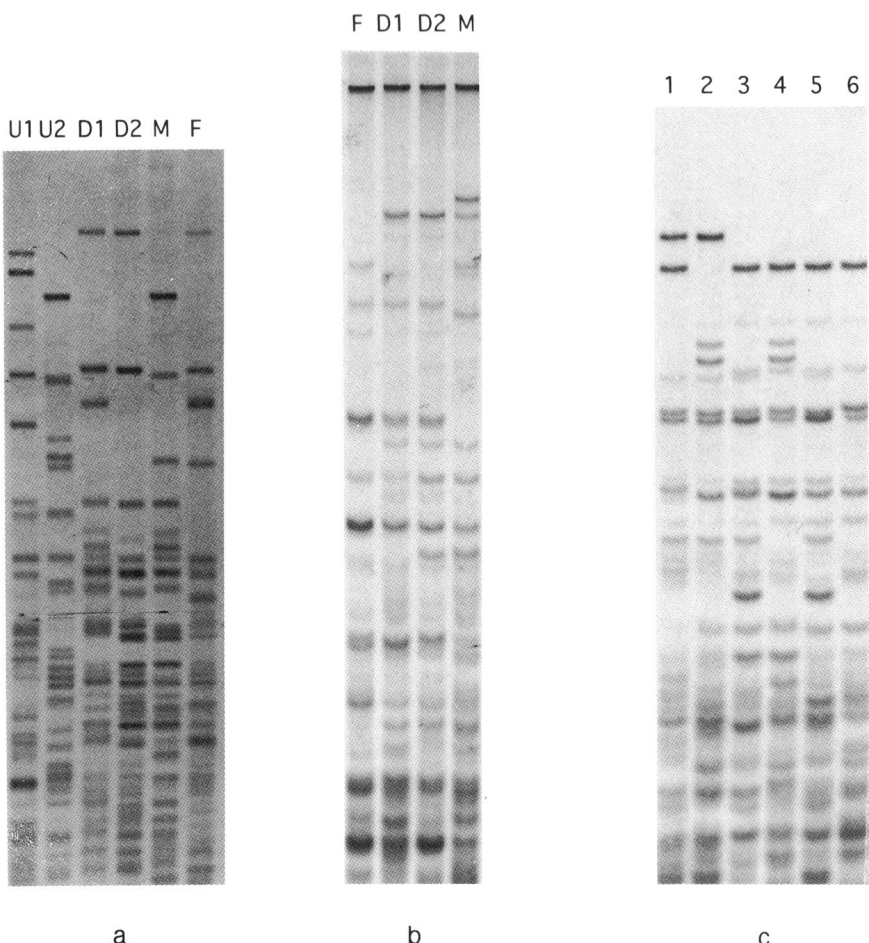

U1 U2 D1 D2 M F

F D1 D2 M

1 2 3 4 5 6

a   b   c

Fig. 1. Examples of DNA fingerprints in three different species using digoxigenin-labeled RNA minisatellite 33.15 and color detection: **(a)** in man, **(b)** in the common marmoset (*Callithrix jacchus*), and **(c)** in the Waldrapp ibis (*Geronticus eremita*). F, father; M, mother; D1 and D2, their daughters; U1, unrelated female; U2, unrelated male; 1 to 6, six related individuals from a captive population. 1.5- to 2 µg DNA digests were loaded per individual. The blots had been hybridized before to at least two other probes.

and the blots prehybridized for at least 8 h. Furthermore, during hybridization and all subsequent processing the blots should not become dry but always be evenly submerged in the solutions.

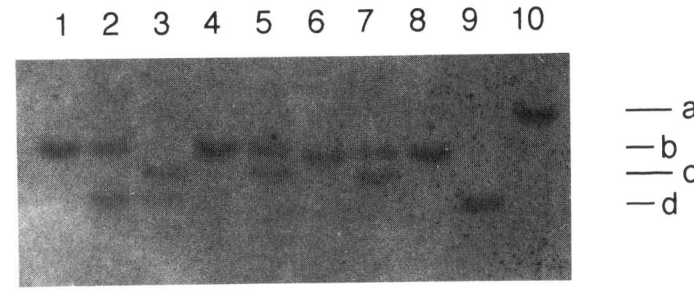

Fig. 2. DNA profiles in eight domestic pigs (1–8) and two wild boars (9 and 10) using digoxigenin labeling and color detection of a porcine minisatellite isolated from a genomic pig library. **a** to **d** indicate the alleles detected at this locus. Each lane contains approx 2 µg of digested DNA.

The visualization of hybridization between the RNA probe and the DNA fragments on the blot is based on an enzyme-linked immunoassay. An antidigoxigenin/alkaline phosphatase conjugate is bound to the digoxigenin component of the probe and then incubated with the substrates X-phosphate and nitroblue tetrazolium (*see* Section 2.) under alkaline conditions, which will result in purple-blue precipitates on the membrane within a couple of hours. Color detection thus saves time and money as it does not require X-ray films, cassettes, and intensifier screens. For a permanent record the result can be photocopied or photographed.

The method described here is a slightly modified version from that suggested by Boehringer *(3)*. As for the labeling of the probe, wearing gloves throughout the whole procedure is imperative. If in doubt about the water quality it should be checked for RNase activity.

2. Materials

The reagents used for the color detection (marked *) may be purchased from Boehringer (Indianapolis, IN). If not indicated otherwise, the amounts given are calculated for one hybridization/color detection experiment using two 20 × 20 cm blots.

1. Digoxigenin-labeled RNA probe (*see* Chapter 12).
2. Southern blot(s) containing 1 µg DNA per lane or more (*see* Note 1).

3. 20X SSC: 3*M* NaCl, 0.3*M* trisodium citrate, pH 7.0. Adjust the pH with concentrated HCl (14 drops from a Pasteur glass pipet per 5 L). Store this solution at room temperature and dilute with distilled water to give 3X SSC, and so on.
4. 3X SSC.
5. 2X SSC.
6. 1X SSC.
7. 20% (w/v) SDS. Do not breathe in the powder. Weigh out in a fume-hood or wear a face mask. The powder will dissolve better when slowly added to preheated distilled water (50°C) on a magnetic stirrer. Store at room temperature. If precipitates form reheat the solution to dissolve. 500 mL will last for several months.
8. Hybridization solution: Work in a fume-hood. To a glass bottle containing about 400 mL distilled water add: 25 mL 20X SSC, 25 mL 20% (w/v) SDS, 5 g dried skimmed milk (e.g., Marvel), 0.1 g sodium azide (**Caution**: contact with water liberates toxic gas), 50 µL diethyl pyrocarbonate (DEPC; **Caution**: suspected carcinogen). Add distilled water to 500 mL. Close the cap firmly and mix thoroughly. Incubate overnight at 42°C with the cap slightly open. Do not autoclave. Store at 4°C. This volume will last for about 10 hybridizations (20 blots).
9. Washing solution 1 (low stringency): 2X SSC, 0.1% (w/v) SDS.
10. Washing solution 2 (intermediate stringency): 1X SSC, 0.1% (w/v) SDS.
11. Washing solution 3 (high stringency): 0.1X SSC, 0.01 % (w/v) SDS. This solution is only required when using single-locus probes that hybridize strongly to other loci.
12. TN buffer: 0.1*M* Tris-HCl, 0.15*M* NaCl, pH 7.5. Prepare 2 L. Autoclave and store at 4°C.
13. Blocking buffer: 0.5% (w/v) blocking reagent* in 400 mL TN buffer. Stir the blocking reagent into the buffer at 60°C. It takes about 1 h to dissolve completely. Do not autoclave. Cool to room temperature before use.
14. ABC (antibody-conjugate) solution: Dilute antidigoxigenin/alkaline phosphatase conjugate* (polyclonal sheep antidigoxigenin F_{ab}-fragments) to 150 mU/mL in 100 mL TN buffer. Prepare on the day of use. This dilution is stable for 12 h (at 4°C). Bring to room temperature before use.
15. Alkaline TNM buffer: 0.1*M* Tris-HCl, 0.1*M* NaCl, 0.05*M* $MgCl_2$, pH 9.5. Autoclave and store at 4°C.
16. NBT solution*: 75 mg/mL nitroblue tetrazolium salt in 70% dimethyl-formamide; store at −20°C.
17. X-phosphate solution*: 50 mg/mL 5-bromo-4-chloro-3-indolyl phosphate, toluidinium salt, in dimethylformamide; store at −20°C.
18. Color solution: Add 112 µL NBT solution and 87 µL X-phosphate

solution to 30 mL alkaline TNM buffer. Prepare immediately before use (*see* Note 2).

19. TE buffer: 0.01*M* Tris-HCl, 0.001*M* EDTA, pH 8.0. Autoclave and store at 4°C.
20. Dimethylformamide (**Caution**: suspected mutagen and teratogen).
21. 0.4*M* NaOH: This solution is conveniently prepared by dilution of an 8*M* stock. Preheat to 50°C before use.
22. 5X Neutralization solution: 1*M* Tris-HCl, 5X SSC, 0.5% (w/v) SDS, pH 7.5. Do not autoclave. Store at room temperature; if precipitate forms, heat the solution to dissolve.
23. 1X Neutralizing solution: Dilute 5X neutralizing solution fivefold in distilled water and preheat to 50°C before use.
24. Hybridization oven with rotor and cylindric glass bottles (30 × 4 cm); nylon meshes (Hybaid, Woodbridge, NJ) and freezer boxes, large enough for 20 × 20 cm blots (*see* Note 3).
25. Transparent file pockets, A4 size (as for stationery use), cut open on three sides.
26. Plastic bag heat sealer (as for domestic use).
27. Cover slip forceps.

3. Methods

3.1. Prehybridization, Hybridization, and Washing

1. Preheat the hybridization solution, hybridization bottles, and oven to 65°C.
2. Load one or two 20 × 20 cm blots for each bottle as follows: In a large tray containing 3X SSC prewet alternately meshes and membranes on top of each other so that one mesh is at the bottom (this will be in contact with the bottle wall) and one between the membranes. Then roll this stack to a cylinder with a smaller diameter than the bottle. Place about 20 mL of 3X SSC into the bottle and push the mesh/membrane roll into the middle of the bottle. Close the lid and hold the bottle horizontally. Turn it slowly so that the membranes and meshes inside start to unroll and make tight contact with the glass wall. When unrolled completely, pour off the 3X SSC and add 30 mL hybridization solution. Firmly close the lid and clamp the bottle onto the oven rotor. Make sure that the membranes remain attached to the bottle wall during rotation. If not, change the orientation of the bottle by 180°. Prehybridize at 65°C and medium rotor speed (10 rpm) for at least 8 h or overnight (*see* Note 4).
3. Thaw the labeled probe on ice. For each hybridization bottle prepare in a sterile tube 15–20 mL hybridization solution with 15 µL probe (use tips from a freshly opened bag). Mix briefly by tilting the tube. Replace the

solution in the bottle with this mixture and immediately start the hybridization at 65°C and medium rotor speed for 16 h (*see* Notes 5 and 6).

4. Preheat the washing solutions 1 and 2 (and, if necessary, solution 3, *see* Note 7) to 65°C (approx 1 L each).

5. Pour off the hybridization solution and add about 50 mL washing solution 1. Put the bottle back to the oven for 5 min but in reverse orientation to loosen the membranes from the wall.

6. Transfer the blots into a freezer box containing washing solution 1 (65°C). Remove the meshes (*see* Note 8) and wash the blots 3 x 20 min in a shaking water bath at 65°C (*see* Note 9).

7. For single-locus probing wash the blots twice, for 15 min each, in washing solution 2 as before. This step is omitted for multilocus probing (DNA fingerprinting) (*see* Note 7).

8. Rinse the membranes twice for 5 min in 1X SSC at room temperature to remove any traces of SDS.

3.2. Color Detection

All the following steps are performed at room temperature with shaking in a sufficient volume to allow the blots to float freely (yet the volumes can be kept low by using a box that is just slightly larger than the blots). Exceptions are indicated. A dot blot containing aliquots of the labeled probe can be included as a control.

1. Wash the blots for 1 min in TN buffer.
2. Incubate twice for 15 min in 200 mL blocking buffer.
3. Wash again for 1 min in TN buffer.
4. Incubate the filters for 30 min in 100 mL ABC solution.
5. Remove the unbound antibody by two washings in TN buffer for 15 min each.
6. Equilibrate the filters for 2 min in alkaline TNM buffer.
7. Seal each membrane separately into an A4 file pocket but leave one side open. Lay the pocket on a tilted support with the open side on top. Add 15 mL of color solution per 20 × 20 cm filter. Remove any air bubbles by slowly rolling a glass pipet over the pocket toward the open side, from time to time releasing the pressure to let the solution flow back. Seal the pocket completely and incubate between two glass plates at room temperature in the dark (cupboard or drawer) for about 8 h (single-locus probing) or overnight (DNA fingerprinting). Do not shake (*see* Notes 10 and 11).
8. Stop the color reaction by submerging the blots in TE buffer. Photocopy or photograph the membranes while they are still wet (the color will fade when the blots dry out).

9. For reprobing destain the membranes in a few mL of dimethylformamide at 65°C (*4*; collect the waste and dispose according to local safety regulations; *see* Note 12). Wash the probe off at 50°C by shaking in two changes of 0.4*M* NaOH for 5 min each followed by two 5 min washes in neutralization solution. Either store the blots in 2X SSC at 4°C or use them for further hybridizations as described.

4. Notes

1. We have used Hybond N FP nylon membranes (Amersham Int., Amersham, UK) extensively. Other types of membranes, however, might work as well. After hybridization, store the blots wet in a closed plastic box containing 2X SSC at 4°C (renew the solution once a month). They will remain in good condition for reprobing over several years (moist in a drawer they would easily become contaminated by bacteria and fungi that destroy the DNA).
2. Boehringer recommend 45 µL NBT plus 35 µL X-phosphate in 10 mL buffer for a single 10×10 cm blot. However, the reduced concentrations applied in the present protocol for two 20×20 cm blots produce good results, too.
3. Perspex chambers or heat resistant bags in a shaking water bath can be used instead. However, these will need larger volumes during prehybridization and hybridization and/or the blots may have to be cut into smaller pieces.
4. Whether more blots (with meshes between them) per bottle would still give a good signal-to-background ratio has not been tested.
5. Do not pipet the probe directly into the bottle as this would cause spots of high background. If a large amount of the probe is applied undiluted it will bind aspecifically to the membranes and cannot be washed off completely afterwards.
6. The amount of probe used per hybridization is based on previous experiments using radioactively labeled RNA probes, where 15 µL probe corresponded to $1\text{-}6 \times 10^7$ cpm.
7. Perform two high stringency washes for 10 min each in washing solution 3 at 65°C when using single-locus probes that tend to crosshybridize strongly to other loci.
8. The meshes can be reused. Wash them in diluted detergent (such as Decon 90) and rinse several times in distilled water (do not crumple them).
9. You may do all posthybridization washes in the bottles, but the efficiency of removing unbound probe, and thus background staining, seems to be better when the blots are washed in larger volumes.
10. The incubation time may vary depending on the amount of DNA on the membrane, the nature of the probe, and the species.

11. If a low signal-to-background ratio is encountered despite the recommendations given, one or more of the following suggestions might help: Wash more extensively after hybridization (longer and/or at higher stringency), increase the concentration of the blocking buffer, add more probe, or use another batch or type of membrane.

12. The flash point of dimethylformamide is 68°C. It is also highly toxic. Therefore, extreme caution should be taken when carrying out this procedure. Use a fume-hood and wear suitable protective clothing and faceshield.

References

1. Signer, E. N. and Jeffreys, A. J. (1992) Both "hot" and "cold" transcripts of minisatellites 33.15 and 33.6 produce informative DNA fingerprints in pigs. *Fingerprint News* **4/2,** 3–7.
2. Carter, R. E., Wetton, J. H., and Parkin, D. T. (1989) Improved genetic fingerprinting using RNA probes. *Nucleic Acids Res.* **17,** 5867.
3. Boehringer Mannheim UK. Nonradioactive DNA Labeling and Detection Kit (No. 1093 657).
4. Gebeyehu, G., Rao, P. Y., SooChan, P., Simms, D. A., and Klevan, L. (1987) Novel biotinylated nucleotide—analogs for labeling and colorimetric detection of DNA. *Nucleic Acids Res.* **15,** 4513–4534.

CHAPTER 19

Hybridization and Detection of Digoxigenin Probes on RNA Blots

Elizabeth Davies, Rachel Hodge, and Peter G. Isaac

1. Introduction

The probing of RNA gel blots (also called Northern blots) with labeled nucleic acids provides data on the relative levels of steady state gene expression, and on RNA processing. The principle of the technique was first described by Alwine et al. *(1)*. The protocol described in this chapter demonstrates the use of nonradioactive digoxigenin-labeled probes (Chapters 10 and 11) on RNA blots (Chapters 7 and 8). The detection system employs an antidigoxigenin antibody/alkaline phosphatase conjugate, and the chemiluminescent substrate 3-(2'-spiroadamantane)-4-methoxy-4-(3"-phosphoryloxy)-phenyl-1,2-dioxetane (AMPPD) *(2,3)*. The procedure used is similar to that described for probing Southern blots and detection by chemiluminescence in Chapter 17. However, this procedure uses a different membrane for support of the immobilized nucleic acids. This, and the fact that RNA is being probed, require modifications to the hybridization buffer and washing procedure.

2. Materials

1. A filter membrane to which RNA has been transferred (*see* Chapters 7 and 8). In the example used in this chapter Hybond N is used (*see* Note 1). The membrane should be handled only at the corners with cover slip forceps.
2. Digoxigenin-labeled probe (*see* Chapters 10 and 11).

From: *Methods in Molecular Biology, Vol. 28:*
Protocols for Nucleic Acid Analysis by Nonradioactive Probes
Edited by: P. G. Isaac Copyright ©1994 Humana Press Inc., Totowa, NJ

3. 20X SSC: 3M sodium chloride, 0.3M trisodium citrate. Adjust to pH 7 with HCl. Filter and autoclave before storing at room temperature.

4. 10 mg/mL herring sperm DNA: This should be denatured by boiling for 10 min, then quenched in ice water before making up hybridization buffer.

5. Hybridization buffer: This should be made up on the day of use. For 100 mL, mix 24 mL of sterile distilled water, 25 mL of 20X SSC, 0.2 g sodium dodecyl sulfate (SDS), and 0.05 g lauryl sarcosine. Heat to 65°C on a heater stirrer and add 0.5 g blocking agent (from Boehringer, Mannheim, Germany). When the blocking agent has dissolved, allow the solution to cool. Immediately before use add 1 mL of 10 mg/mL freshly denatured herring sperm DNA and 50 mL formamide (analytical reagent grade, Fisons, Loughborough, UK) (**Caution:** This is a teratogen, and highly toxic. Use in a fume cupboard).

6. Wash A: 2X SSC, 0.1% SDS. Autoclave—this solution may be stored at 37°C to prevent SDS precipitation.

7. Wash B: 0.1X SSC, 0.1% SDS. Autoclave and store at 42°C.

8. Buffer 1: 100 mM Tris-HCl, pH 7.5, 150 mM NaCl. Filter and autoclave.

9. Buffer 2: 0.5% Boehringer blocking agent dissolved in autoclaved and hot (about 60°C) buffer 1 (this takes some time to dissolve, it is usually best to dissolve this on a stirring hot-plate) (*see* Note 2).

10. Buffer 3: 100 mM Tris-HCl, pH 9.5, 100 mM NaCl, 50 mM MgCl$_2$. This should be prepared fresh on the day of use. Filter but do not autoclave. Check the pH of this solution immediately before use (Section 3.2., step 8).

11. Antidigoxigenin antibody/alkaline phosphatase conjugate (available from Boehringer): A 1/5000 dilution in Buffer 1 should be made immediately before use (Section 3.2., step 4). 30 mL are required for one 10 × 18 cm filter, 70 mL for 10 filters.

12. AMPPD (from Tropix Inc., Bedford, MA or Boehringer): This is normally supplied as a 10 mg/mL stock solution and is diluted before use. Add 10.6 µL of 10 mg/mL AMPPD/mL of buffer 3. The procedure requires 20 mL for a filter of 10 × 18 cm, and 50 mL for 10 filters (*see* Note 3).

3. Methods

3.1. Prehybridization, Hybridization, and Washing

1. Place the filter in a plastic bag, add 30 mL (for a single 18 × 8 cm filter) to 100 mL (for 10 filters of the same size) prehybridization solution. Remove any air bubbles and seal the bag. Prehybridize for 2 h at 42°C (shaking is not necessary if a single blot is being analyzed).

2. Place a further aliquot of the prehybridization solution in the oven at 42°C for this time.

3. Boil the probe for 10 min (about 150 ng for one filter, use 300 ng for 10 filters).

4. Quench the probe in iced water for 5 min (*see* Note 4) and then add this to the unused prehybridizing solution at 42°C and mix thoroughly.
5. Tip out the prehybridizing solution and add the probe solution (15 mL for one 18 × 8 filter, 40 mL for 10 filters).
6. Hybridize 16 h at 42°C (shaking is not necessary if a single blot is being analyzed). Meanwhile place Wash B in the 42°C incubator.
7. Collect the probe into clean, sterile container, and store at –20°C for further reuse.
8. Transfer the filters into a clean sandwich box and wash the filters in two changes of 5 min in Wash A, nominally at room temperature (the solution is kept stored at 37°C to prevent SDS precipitation and so is at 37°C at the start of the wash).
9. Wash filters in two changes of 15 min in Wash B at 42°C with shaking.

3.2. Detection of Probe

The main washing steps are done with the filters loose in a box in a volume of 300 mL for a single filter, 400 mL for a group of 10 filters, with shaking. The incubation in diluted antibody, and the incubation with AMPPD substrate are done in a plastic bag with shaking. All these steps are nominally at room temperature

1. Rinse the filters for 1 min in sterile buffer 1.
2. Block the filters by washing for 30 min in buffer 2.
3. Rinse for 1 min in sterile buffer 1.
4. Transfer the filters to a plastic bag, add 1/5000 dilution of antibody conjugate in buffer 1. Seal the bag and shake for 30 min.
5. Transfer the filters back to a box. Wash for two 15-min changes with shaking in buffer 1.
6. Wash for 30 min (*see* Note 5), and then three 5-min changes, in buffer 2. This is to block the membrane again before incubation in AMPPD.
7. Wash for four 5 min changes in buffer 1.
8. Wash for 5 min in buffer 3. This wash is necessary to raise the pH of the filters.
9. Transfer the filters to a plastic bag, and add diluted AMPPD. Shake for 5 min.
10. Pour off AMPPD into a clean sterile container, wrap in tin foil, and keep at 4°C (*see* Note 3). Do not blot the filters dry. Wrap filters in Saran Wrap, RNA side up.
11. Expose at room temperature against X-ray film. Usually 10 min to check the background, then a full exposure for 2 h or so should show low level transcripts. Longer exposures can also be done (4 h to overnight) to bring up faint tracks if the background is sufficiently low (*see* Note 6).

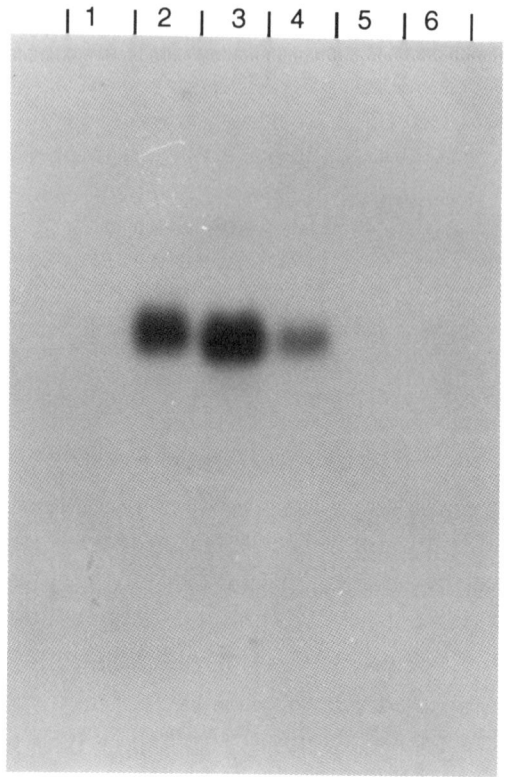

Fig. 1. RNA gel blot of total RNA isolated from *Brassica napus* buds probed with the Arabidopsis A9 gene *(4)*. The probe was labeled with digoxigenin by the polymerase chain reaction (Chapter 10). The RNA samples on the gel are bud size <1 mm (lane 1), 1–2 mm (lane 2), 2–3 mm (lane 3), 3–4 mm (lane 4), 4–5 mm (lane 5), >5 mm (lane 5). Each lane contains 10 µg of total RNA. The exposure time was 30 min.

12. The result should be dark bands superimposed on a clear background e.g., Fig. 1 (*see* Notes 7–9).

4. Notes

1. Hybond N membrane differs from that used in Chapter 4 in that it is less highly charged. This membrane should not be used with high concentrations of SDS. If a more highly charged membrane is used, e.g., Hybond N+, or Pall Biodyne B, then the level of SDS should be increased.
2. Skimmed milk powder can replace blocking agent in this buffer. We have used Marvel at a final concentration of 1%. Marvel milk powder, surprisingly, should be stored frozen.

3. Diluted AMPPD can be reused at least three times.

4. If using a recycled probe in a large volume of solution, then give this longer to cool down, e.g., about 15 min. Recycled probes should be heated in a fume-hood. Because of the presence of formamide they will denature at 70°C.

5. We have left the filters for longer at this stage (up to 2 h) without ill effect.

6. If it is not possible to do a long exposure on the day the filters are developed, they may be stored in the Saran Wrap at –20°C overnight, warmed up the next day and then exposed against film. The rate of signal production plateaus 12–16 h after exposure to AMPPD and remains constant for several days, hence a 10-min exposure of filters after a delay of 24 h will produce a signal strength 2–4 times that produced on the first day. One of the benefits of this system is that duplicate copies of luminographs can be produced for many days after the original probing.

7. If the film is covered with amorphous black smudges, these are likely to have been caused by poor handling of the membrane. These marks cannot be stripped off easily.

8. If the film has a uniformly high background, then the blocking washes are likely to have been inadequate, or the membrane used is highly charged and requires the addition of more SDS (up to 0.5%) in the prehybridization solution.

9. If the background of the filter is good, but the tracks are uniformly black, then washing at a higher stringency may be necessary. This problem can sometimes be avoided by including two washes in the hybridization buffer at 42°C after Section 3.1., step 9. Other explanations can be poor removal of proteins from samples or degraded RNA. It is usually best to perform fresh RNA extractions if this is the case.

References

1. Alwine, J. C., Kemp, D. J., and Stark, G. R. (1977) Method for detection of specific RNA in agarose gels by transfer to diazo benzyloxymethyl paper and hybridization with DNA probes. *Proc. Natl. Acad. Sci. USA* **74,** 5350–5354.

2. Voyta, J. C., Edwards, B., and Bronstein, I. (1988) Ultrasensitive chemiluminescent detection of alkaline phosphatase activity. *Clin. Chem.* **34 ,**1157.

3. Bronstein, I., Voyta, J. C., and Edwards, B. (1989) A comparison of chemiluminescent and colorimetric substrates in a hepatitis B virus DNA hybridization assay. *Analyt. Biochem.* **180,** 95–98.

4. Paul, W., Hodge, R., Smartt, S., Draper, J., and Scott, R. (1992) The isolation and characterisation of the tapetum-specific Arabidopsis A9 gene. *Plant Mol. Biol.* **19,** 611–622.

CHAPTER 20

Hybridization of Horseradish Peroxidase-Labeled Probes and Detection by Enhanced Chemiluminescence

Timothy Stone and Ian Durrant

1. Introduction

The use of probes directly labeled with horseradish peroxidase (HRP) in conjunction with enhanced chemiluminescence (ECL) allows a flexible approach to hybridizations and detections. This is shown by the diversity of applications in which this labeling and detection system has been used. The applications include Southern blots *(1,2)* and Northern blots *(3)*, colony and plaque screening for positive clones *(4)*, YAC clone screening *(5)*, and PCR product detection *(6,7)*.

The major steps required for the use of directly labeled HRP probes are hybridization, stringent washes, and detection.

1. Probes labeled directly with HRP are hybridized in 6*M* urea, a denaturing buffer. The maximum rate of hybridization occurs at 25°C below T_m *(8)*. Since the T_m of most probes is approx 67°C in 6*M* urea *(9)*, hybridizations are optimal at 42°C. These conditions are equivalent to hybridizing a radiolabeled probe at 65°C in a nondenaturing buffer, or hybridizing at 42°C in 50% formamide. HRP is stable at 42°C for at least 24 h, so that overnight hybridizations can be carried out to gain maximum sensitivity. The composition of the buffer has been optimized *(10)* and contains a novel rate enhancer and blocking agent, which are

From: *Methods in Molecular Biology, Vol. 28:*
Protocols for Nucleic Acid Analysis by Nonradioactive Probes
Edited by: P. G. Isaac Copyright ©1994 Humana Press Inc., Totowa, NJ

both necessary to prevent the labeled probe binding nonspecifically to nylon or nitrocellulose membranes.

2. Stringency can be controlled, just as with radiolabeled probes, using the salt (SSC) concentration in the primary wash buffer. These stringent washes take <60 min in total. There are no time-consuming antibody blocking or incubation steps, making detection with directly labeled HRP probes ideal for large-scale screening, where many blots are handled simultaneously.

3. Detection by ECL involves the oxidation of luminol that is catalyzed by the HRP on the probe. The oxidized luminol decays to the ground state via a light emitting pathway. In the presence of an enhancer molecule there is at least 1000-fold more light produced *(11)*. The wavelength of maximum emission occurs at 428 nm, allowing detection on blue-light sensitive film. The light output peaks after 1–5 min, and then decays slowly with a 2-h half-life *(10)* making multiple or extended exposures possible. Once the light output has decayed, owing largely to irreversible HRP inactivation, blots can be immediately reprobed, without stripping. Detection by ECL gives fast and sensitive results that are recorded as a hard copy on film with excellent resolution. The use of probes labeled directly with HRP, in conjunction with ECL detection, provides a flexible and sensitive system for identifying any nucleic acid of interest. The probe concentration, length of hybridization, and the exposure time can be altered to suit the needs of a particular application.

2. Materials

The hybridization buffer, blocking agent, and detection reagents are available in an optimized kit form, including the reagents for labeling probes directly with HRP. (ECL direct nucleic acid labeling and detection system, RPN 3000, RPN 3001, RPN 3005, Amersham International, Amersham, UK).

1. Hybridization buffer and blocking agent (Amersham): The hybridization buffer is a complex mixture of components that have been optimized to ensure efficient hybridization and thermal stability of the HRP label. It is based on $6M$ urea as a denaturant and also contains a novel rate enhancement system. This system causes greater rates of hybridization and also aids in the stabilization of the HRP during the detection process. At room temperature, take the hybridization buffer and add sufficient solid NaCl to make a $0.5M$ final concentration. Add 5% (w/v) blocking agent (this is necessary for both nylon and nitrocellulose membranes). Quickly mix the blocking agent into the buffer to ensure that no clumps of undissolved blocking agent form. Mix thoroughly, using

a magnetic stirrer or roller mixer, for 1 h. The blocking agent is not fully dissolved at this stage, and heating to 42°C for 30–60 min is necessary to complete the process. A few undissolved particles of blocking agent will not affect the hybridization.

Hybridization buffer and blocking agent, if stored separately, are stable at room temperature. However, when formulated as above, the hybridization buffer must be stored at –20°C, where it is stable for at least 6 mo.

2. 20X SSC: 0.3M trisodium citrate, 3.0M sodium chloride. Adjust to pH 7 with HCl.
3. Primary wash buffer: 360 g Urea, 4 g SDS, 25 mL 20X SSC. Make up to 1 L. This can be kept for up to 3 mo in a refrigerator at 2–8°C. Stringency may be increased by using a lower final SSC concentration, e.g., 0.1X SSC instead of 0.5X SSC.
4. Secondary wash buffer: 2X SSC. This can be kept for up to 3 mo in a refrigerator at 2–8°C.
5. Saran Wrap™ (Dow Chemical Company).
6. Blotting membranes: The system gives excellent results with Hybond™-N⁺ (nylon) and Hybond-ECL (nitrocellulose) membranes. Membranes supplied by other manufacturers may also give good results.
7. Detection reagents 1 and 2 (Amersham): Detection reagent 1 contains a peracid salt as the enzyme substrate. Detection reagent 2 contains luminol and an optimized enhancer molecule. An equal volume of each of these reagents should be mixed. The final volume required for ECL detection is 0.125 mL/cm² of membrane. The detection reagents are stable for at least 3 mo when stored separately at 2–8°C.
8. X-ray film cassette (e.g., Hypercassettes, Amersham).
9. Blue-light sensitive film (e.g., Hyperfilm™-ECL, Amersham).
10. Horseradish peroxidase-labeled probe (*see* Chapter 14): The labeling procedure produces a probe in a single-stranded form; it should not be denatured further.

3. Methods
3.1. Hybridization

1. Place the blot onto the surface of the fully prepared hybridization buffer, prewarmed to 42°C. Allow the blot to saturate fully before submerging in the buffer (*see* Note 1).
2. Prehybridize at 42°C for15–60 min in a shaking water bath (*see* Note 2).
3. Add the single-stranded probe, which has been labeled directly with HRP (*see* Chapter 14), to the hybridization buffer (*see* Note 3), avoiding direct addition onto the membrane (*see* Note 4).

4. Continue incubation at 42°C with shaking for the required length of time (from 2 h to overnight) (*see* Note 3).

3.2. Stringent Washes

1. Preheat the primary wash buffer to the required temperature, typically 42°C (*see* Note 5).
2. Pour at least 2 mL primary wash buffer/cm^2 of membrane into a clean container.
3. Remove the blots from the hybridization container and place them into the primary wash buffer. Incubate at the desired temperature in a shaking water bath for 20 min.
4. Discard the wash buffer and replace with an equivalent volume of the fresh primary wash buffer. Incubate with agitation for a further 20 min at the same temperature.
5. Discard wash buffer, place blots in a fresh container, and add an excess of secondary wash buffer (at least 2 mL of buffer/cm^2 of membrane). Incubate with agitation for 5 min at room temperature.
6. Discard the wash buffer and replace with an equivalent volume of fresh secondary wash buffer. Incubate for a further 5 min at room temperature.
7. The blots may be stored temporarily at this stage if required (*see* Note 6).

3.3. Detection

1. Lay a piece of Saran Wrap, with a large enough area for all the blots, onto the bench.
2. Using forceps remove the blots from the secondary wash buffer. Drain off excess wash buffer by touching the bottom edge of the blot onto a paper towel and place the blots, DNA side up, onto the Saran Wrap.
3. Cover the blots with a volume of freshly mixed detection reagents 1 and 2 equivalent to 0.125 mL/cm^2 of membrane.
4. Leave the blots immersed for 1 min.
5. Drain off excess detection reagents by touching the bottom edge of the blot onto a paper towel and place the blots, DNA side down, onto a fresh piece of Saran Wrap.
6. Fold over the Saran Wrap to make a "parcel" of the blots. Only one layer of Saran Wrap should cover the DNA side of the blots.
7. In a darkroom, with a red safelight, place the blot parcel DNA side up in the base of an X-ray film cassette. Place a piece of blue-light sensitive film on top, close the cassette, and expose the film for 1 min.
8. Remove the film and immediately replace with another sheet of unexposed film, reclose the cassette, and start a timer. Develop the first film and use result as a guide for the length of the second exposure (*see* Note 7).

9. After a suitable interval, normally 30–60 min for genomic Southern blots, develop the second film and interpret results.

4. Notes

1. If hybridization is performed in a box, then the blot must be placed on top of the buffer and allowed to saturate fully before being submerged. This prevents the appearance of white patches on detection. A volume of buffer equivalent to 0.25 mL/cm^2 of membrane should be used. If the box is significantly larger than the blot then the volume used should correspond to the area of the bottom of the box. Hybridizations may also be carried out in bags or in hybridization ovens. In these cases the blot should be put into the container first. Buffer is added at a volume equivalent to 0.125 mL/cm^2 of membrane.

2. Shaking water baths should be set to 60–100 strokes/min. Blots should be allowed to move freely in boxes to prevent the appearance of white or dark patches on detection.

3. Routinely 10 ng/mL final probe concentration and an overnight hybridization is used. This gives the maximum sensitivity of the system and is a requirement for high sensitivity work, for example the detection of a single-copy gene on Southern blots of genomic DNA digests (9). However, for high target applications, such as colony or plaque screening, it may be possible to use lower probe concentrations of (2–5 ng/mL) in conjunction with a 2-h hybridization time *(10)*.

4. Probe added directly onto the membrane may result in high patchy background on detection around the site of the probe addition.

5. Stringency is controlled in the primary washes by the SSC concentration in a similar manner to posthybridization stringency washes with ^{32}P-labeled probes. Routinely the primary wash contains 6*M* urea and is performed at 42°C. However, it is possible to achieve similar results using a primary wash without urea with incubations performed at 55°C. Using a temperature greater than those recommended here, for either system, will result in either a very low signal, because of the stringency being too great or, no signal at all, because of inactivation of the peroxidase. Each of the primary washes must be carried out for no longer than 20 min for similar reasons.

6. Blots may be stored for up to 24 h at this stage, either at 4°C, wetted in secondary wash buffer, and wrapped in Saran Wrap, or at –20°C, immersed in a solution of 50% glycerol. After storage, blots should be rinsed briefly in fresh secondary wash buffer just before they are detected.

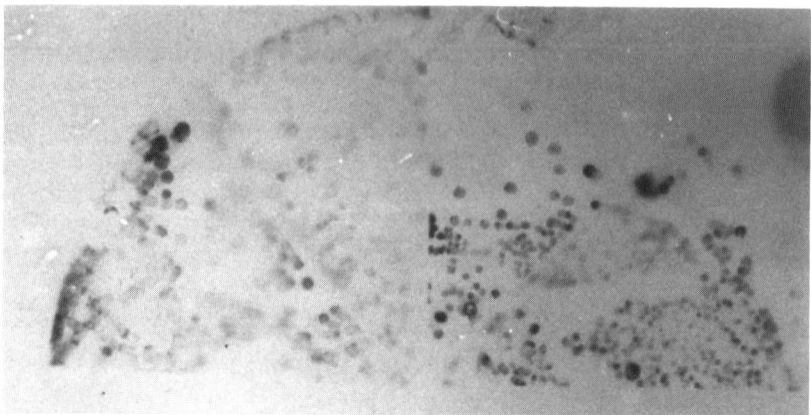

Fig. 1. Colony screening. *E. coli.* colonies transformed with a plasmid carrying a portion of the proto-oncogene Nras hybridized with a 1.5-kbp HRP-labeled Nras probe as described in the text. 1-min exposure on Hyperfilm-MP.

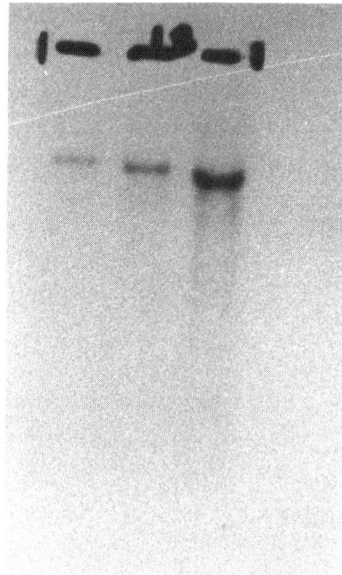

Fig. 2. Single copy gene detection. 1, 2, and 5 µg loadings of an *Eco*RI restriction enzyme digest of human genomic DNA immobilized on Hybond-N$^+$. Hybridized with a 1.5-kbp HRP-labeled Nras proto-oncogene probe as described in the text. 30-min exposure on Hyperfilm-MP.

7. The length of the second exposure can be judged by assessment of the signal and background levels observed on the first film. As a guideline, typical exposure times would be 5–10 min for high target systems (e.g., colony and plaque screening) and 30–60 min for low target levels, such as those found in single copy gene detection (*see* Figs. 1 and 2).

References

1. Heslop-Harrison, J. S. (1990) Gene expression and parental dominance in hybrid plants. *Development* (**suppl.**) 21–28.
2. Tonjes, R. R. (1991) Detecting transgene sequences by Southern blot analysis using the ECL gene detection system. *Life Sci.* **4**, 7,8.
3. Einspanier, R. and Scham, D. (1991) Determination of insulin-like growth factor I and II in bovine ovarian follicles and cysts. *Acta Endocrinol.* **124 (suppl. 1)**, 19.
4. Durrant, I., Benge, L. C. A., Sturrock, C., Devenish, A. T., Howe, R., Roe, S., Moore, M., Scozzafava, G., Proudfoot, L. M. F., Richardson, T. C., and McFarthing, K. G. (1990) The application of enhanced chemiluminescence to membrane based nucleic acid detection. *Biotechniques* **8**, 564–570.
5. Stone, T. and Durrant, I. (1992) Enhanced chemiluminescence for the detection of bound nucleic acid sequences. *Genetic Analysis: Techniques and Applications* **8**, 230–237.
6. Sorg, R., Enczmann, J., Sorg, U., Kogler, G., Schneider, E. M., and Wernet, P. (1990) Specific non-radioactive detection of PCR-amplified sequences with enhanced chemiluminescence labeling. *Life Sci.* **2**, 3,4.
7. Tucker, S. J. (1991) ECL direct system: Identification of the cystic fibrosis gene of *Xenopus laevis*. *Highlights* **2**, 4.
8. Britten, R. J. and Davidson, E. H. (1985) Hybridization strategy in *Nucleic Acid Hybridization: A Practical Approach* (Hames, B. D. and Higgins, S. J., eds.), IRL, Oxford, pp. 3–15.
9. Hutton, J. R. (1977) Renaturation kinetics and thermal stability of DNA in aqueous solutions of formamide and urea. *Nucleic Acids Res.* **4**, 3537–3555.
10. Stone, T. (1992) ECL direct system. An analysis of filter hybridization kinetics. *Life Sci.* **7**, 8.
11. Durrant, I. (1990) Light based detection of biomolecules. *Nature* **346**, 297, 298.

CHAPTER 21

Hybridization and Detection of Fluorescein-Labeled DNA Probes Using Enhanced Chemiluminescence

Bronwen M. Harvey and Claire B. Wheeler

1. Introduction

DNA probes can be labeled using the modified nucleotide, fluo-rescein-11-dUTP, and subsequently detected with enhanced chemilu-minescence *(1)* using an antifluorescein antibody conjugated to horseradish peroxidase *(2)*. The use of enhanced chemiluminescence avoids the fading of results associated with some colormetric detection procedures, used with nonradioactive labeling and detection systems. Hard copy results showing excellent resolution are produced.

The labeling procedure, based on established random primer tech-nology *(3)* can be used with DNA of various types and purity (*see* Chapter 15). The resulting fluorescein-labeled probes can be stored for several months, thus avoiding frequent labeling reactions. Valu-able probe is not lost since a further purification step before hybrid-ization is not required.

The hybridization and wash conditions given in the following pro-tocol are appropriate for a majority of probes labeled using the ECL™ random prime labeling system. They allow the detection of single copy mammalian genes without significant crosshybridization to nonhomologous sequences. However, if these conditions are found to be insufficiently stringent for a particular probe, the stringency can be

From: *Methods in Molecular Biology, Vol. 28:*
Protocols for Nucleic Acid Analysis by Nonradioactive Probes
Edited by: P. G. Isaac Copyright ©1994 Humana Press Inc., Totowa, NJ

further controlled in hybridization and washing by changing the temperature and/or sodium chloride concentration.

2. Materials

Reagents for labeling with fluorescein-11-dUTP and its subsequent detection are available, as a complete system, from Amersham International (Amersham, UK) as ECL random prime labeling and detection system (RPN 3030/3031/3040).

1. Fluorescein-labeled probe DNA (*see* Chapter 15).
2. Hybridization buffer: 5X SSC, 0.5% (w/v) blocking agent (RPN3023, Amersham International), 0.1% (w/v) sodium dodecyl sulfate (SDS), 5% (w/v) dextran sulfate, 100 μg/mL sheared denatured heterologous DNA. Combine all the reagents except the dextran sulfate and the DNA, heat with gentle stirring to 55–60°C. The blocking agent should take 30 min to dissolve leaving a slightly opaque solution with no particulate matter present. Remove from the heat and add the dextran sulfate, a little at a time while stirring, until dissolved, which should take a further 15–30 min. The hybridization buffer can be subaliquoted and stored at –20°C for at least 3 mo (*see* Note 1).
3. Stringent wash solution I: 1X SSC, 0.1% (w/v) SDS.
4. Stringent wash solution II: 0.5X SSC, 0.1% (w/v) SDS.
5. Wash buffer: 100 mM Tris-HCl, pH 7.5, 150 mM NaCl.
6. Antibody blocking buffer: 0.5% (w/v) blocking agent (RPN3023, Amersham International), 100 mM Tris-HCl, pH 7.5, 150 mM NaCl. Dissolve the blocking agent by heating to 55–60°C for approx 30 min. This can be stored at –20°C for several weeks.
7. Antibody binding buffer: 0.5% (w/v) crystalline bovine serum albumin fraction V (e.g., Sigma [St. Louis, MO] A-2153), 100 mM Tris-HCl, pH 7.5, 150 mM NaCl.
8. Antibody wash buffer: 100 mM Tris-HCl, pH 7.5, 150 mM NaCl, 0.1% (v/v) Tween 20.
9. TE buffer: 10 mM Tris-HCl, pH 7.5, 1 mM EDTA.
10. Antifluorescein antibody/horseradish peroxidase conjugate. Available from Amersham International (RPN 3022).
11. ECL detection reagents. Available from Amersham International (RPN 3004/2105). The detection reagent 1 contains the enzyme substrate peroxide. Detection reagent 2 contains an enhancer molecule and luminol, the oxidation of which is linked to the reduction of peroxide.
12. A roll of Saran Wrap™, available from Dow Chemical Company.

13. Blue-light sensitive autoradiography film, e.g., Hyperfilm™-ECL (Amersham International).
14. X-ray film cassette, e.g., Hypercassette™ (Amersham International).

3. Methods

3.1. Hybridization and Stringent Washes

1. Thaw the required volume of hybridization buffer. It is recommended that at least 0.125 mL/cm² of membrane be used (*see* Note 2). Preheat the buffer to 60°C.
2. Complete the preparation of the buffer by adding sheared denatured heterologous DNA to a final concentration of 100 μg/mL (*see* Note 3).
3. Prehybridize the blots in the hybridization buffer for at least 30 min at 60°C with constant agitation (*see* Note 4).
4. Remove the required amount of probe to a clean microcentrifuge tube. If the volume is <20 μL make up to this volume with water or TE buffer. Recommended probe concentrations are 10–20 ng/mL for low target applications and 1–5 ng/mL for high target applications (*see* Note 5). Denature the probe by boiling for 5 min and snap cool on ice.
5. Centrifuge the denatured probe briefly then add to the prehybridization buffer. Avoid placing it directly on the membrane and mix gently. Alternatively, some of the buffer can be withdrawn for mixing with the probe before addition to the bulk of the buffer. Continue the incubation overnight at 60°C (*see* Note 6).
6. Preheat stringent wash solutions I and II to 60°C. These are used in excess, e.g., 2–5 mL/cm² of membrane. Carefully transfer the blots, using a pair of blunt forceps, to stringent solution I and wash at 60°C, for a minimum of 15 min with gentle agitation. Carry out a further wash in stringent wash solution II for 15 min at 60°C (*see* Note 7).
7. Briefly rinse the blots in wash buffer at room temperature to remove any SDS.

3.2. Blocking, Antibody Incubation, and Washes

1. Incubate the blots for 60 min at room temperature, in approx 0.5 mL/cm² of antibody blocking buffer with gentle agitation (*see* Note 8).
2. Dilute the antifluorescein/horseradish peroxidase conjugate 1000-fold in freshly prepared antibody binding buffer. The final volume should be at least equivalent to that used for hybridization. Incubate the blots in diluted conjugate with gentle agitation at room temperature for 60 min (*see* Note 9).

3. Remove unbound conjugate by washing for 2×10 min followed by 2×5 min in antibody wash buffer at room temperature with agitation. An excess volume should be used ($2–5$ mL/cm^2).

3.3. Signal Generation and Detection

Please read through the section before proceeding. For optimum sensitivity it is necessary to work with reasonable speed once the blots have been exposed to the detection solution, as there is no lag phase with ECL detection. All steps can be carried out in the dark-room if desired: It is only necessary to switch off the light after step 4. Avoid using powdered gloves as the powder can cause blank patches on the film.

1. Mix an equal volume of detection solution 1 with detection solution 2 to give sufficient volume to cover the blot (0.125 mL/cm^2).
2. Drain the excess buffer from the washed blots and place them on a sheet of Saran Wrap DNA side upward.
3. Add the detection solution directly to the blot(s). Incubate for 1 min at room temperature.
4. Drain off excess detection buffer by touching the edge of the blot to a tissue, and wrap the blots in fresh Saran Wrap.
5. Place the blots DNA side up in the film cassette. Place a sheet of auto-radiography film on top of the blots. Close the cassette (*see* Note 10).
6. For Southern blot applications, such as the detection of single copy mammalian genes, an exposure of 60 min (*see* Note 11) should give a suitable image (*see* Fig. 1). Shorter exposures should be used for high target application, such as colony and plaque lifts or the detection of PCR amplified target. Film exposure times of $0.5–10$ min are suggested. It is important to use the minimum exposure time that enables positive signal to be unambiguously detected.

4. Notes

1. The hybridization buffer may be used for most membrane-based hybrid-ization processes. However, for Northern blot applications the follow-ing modification to the hybridization buffer is required. The blocking agent must be autoclaved before it can be used. Dissolve the blocking agent in 100 mM trisodium citrate, 150 mM NaCl to give a 10X solu-tion, that is 5 g/100 mL. Autoclave for 15 min at 15 lb/in.^{2} on liquid cycle. The autoclaved solution will have a slight pink coloration.
2. Hybridization can be carried out in either boxes or bags provided there is sufficient buffer. As a general rule 0.25 mL/cm^2 is recommended for small blots and 0.125 mL/cm^2 for large blots. It is possible to hybridize

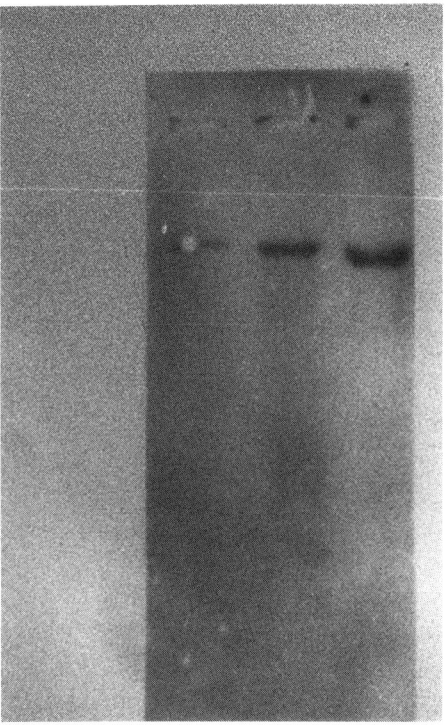

Fig. 1. Single copy gene detection. 1, 2, and 5 µg of an *Eco*RI digest of human genomic DNA blotted onto Hybond-N⁺. Hybridized with a 1.5-kb probe for N-ras proto-oncogene labeled with fluorescein-dUTP by the random primer labeling reaction (*see* Chapter 15). Hybridization with 10 ng/mL of probe, overnight, at 60°C; 30-min exposure on Hyperfilm-MP.

    several blots in the same solution providing there is enough buffer to allow free movement of the blots. Agitation during hybridization is strongly recommended; a speed of 60–100 strokes/min is adequate.

3. The addition of sheared, denatured heterologous DNA to the hybridization buffer has been found to reduce nonspecific hybridization, and it has therefore been included as a standard component. However, with probes showing a low level of crosshybridization to other sequences in genomic DNA it is possible to omit this component. This can result in increased specific signal.

4. Some particulate material may be found in the buffer after thawing but this should redissolve during hybridization at 60°C. When hybridizing large blots prewetting in 5X SSC is advised to minimize the hybridization volume.

5. For hybridization to Southern blots of human genomic DNA, a probe concentration of 10 ng/mL of labeled purified plasmid insert is usually sufficient to detect a single copy gene in a 1–2 μg loading: representing 0.5–1.0 attamol of specific target. If an intact plasmid probe is to be used then it is necessary to correct for the proportion of plasmid sequence present; an overall probe concentration of 20 ng/mL is likely to be required. The signal:noise ratio of the final result is sometimes lower than that obtained with insert alone.

6. It is possible to use shorter hybridization times, although some loss of sensitivity will result. This can be offset by the use of higher probe concentrations.

7. To increase the stringency, washes of 0.2 or 0.1X SSC with 0.1% (w/v) SDS at 60°C can be used. Alternatively, the hybridization and/or stringent washes can be carried out at 65°C.

8. This is a critical step in the detection procedure. Efficient blocking of the membrane against nonspecific binding of the antibody conjugate is essential in minimizing background. This is particularly important in colony/plaque screening. The blocking agent supplied has been specially selected for its solubility and superior blocking capabilities. An excess of antibody blocking buffer, e.g., 1–2 mL/cm² of membrane is advised.

9. With high target applications it is possible to dilute the antibody further; e.g.,1 in 2500.

10. The enzymatic reduction of peroxide is coupled to the oxidation of luminol. As the luminol breaks down it passes through an excited state and light is emitted. This light output (λ_{max} 428 nm) is detected on blue-sensitive film providing a permanent, hard copy result.

11. A second exposure of 60–120 min will give an equivalent image to a first exposure of 30–60 min. To maximize the signal the film can be preflashed. This procedure also raises background, but can be a useful approach if overall signal is low. A single flash from an optimized preflash unit (e.g., Sensitize™ RPN2051, Amersham International), positioned to give an OD of 0.1 absorbance U above that of unexposed film is recommended.

References

1. Whitehead, T. P., Thorpe, G. H. G., Carter, C. J. N., Groucutt, C., and Kricka, L. J. (1983) Enhanced luminescence procedure for sensitive determination of peroxidase labeled conjugates in immunoassay. *Nature* **305,** 158,159.

2. Simmonds, A. C., Cunningham, M., Durrant, I., Fowler, S. J., and Evans, M. R. (1991) Enhanced chemiluminescence in filter based DNA detection. *Clin. Chem.* **37,** 1527,1528.

3. Feinberg, A. P. and Vogelstein, B. (1983) A technique for radiolabeling DNA restriction endonuclease fragments to high specfic activity. *Anal. Biochem.* **132,** 6–13.

CHAPTER 22

Hybridization of Fluorescein-Labeled Oligonucleotide Probes and Enhanced Chemiluminescence Detection

Ian Durrant and Patricia M. E. Chadwick

1. Introduction

Oligonucleotide synthesizers are now commonplace with the result that oligonucleotides can be obtained readily, obviating the need for complex molecular biological techniques. However, the choice between long probes or oligomers depends, to a large extent, on the application; oligonucleotide probes do offer several advantages over long DNA probes in membrane hybridizations.

For example, the specificity of an oligonucleotide probe can be chosen so that single base changes in target sequences can be recognized, since a single-base mismatch in a short probe will decrease the T_m of the hybrid by approx 5°C *(1)*. The short length and low mol wt of oligonucleotides means that hybridization rates are more rapid; in a typical experiment 1–2 h would probably be adequate. The short length of the oligonucleotide may mean a less specific probe when compared with a cloned probe, but this can be averted by careful selection of sequences and by extending the length of the oligonucleotide beyond 30 bases. The following method describes the use of fluorescein-labeled oligonucleotides on membranes. The technique is rapid, simple, and nonradioactive, with detection of hybrids by enhanced chemiluminescence.

From: *Methods in Molecular Biology, Vol. 28:*
Protocols for Nucleic Acid Analysis by Nonradioactive Probes
Edited by: P. G. Isaac Copyright ©1994 Humana Press Inc., Totowa, NJ

A simple hybridization buffer is recommended and hybridization time may range from 2 h to overnight, as convenient. A probe concentration of 5–10 ng/mL is usually sufficient and both nitrocellulose and nylon membranes may be used. The best results, however, will be obtained with charged nylon membranes. The hybridization stringency of the oligonucleotide probe is unaffected by the presence of the tail *(2)* synthesized under standard labeling conditions (*see* Chapter 16). Thus, stringency can be controlled in the usual way, by increasing the temperature or decreasing the salt concentration *(3)*.

Detection of hybridized oligonucleotides is a two-step process. The fluorescein hapten is first bound to an antifluorescein antibody/horseradish peroxidase (HRP) conjugate. The bound peroxidase is then detected using enhanced chemiluminescence (ECL). In this process, the horseradish peroxidase catalyzed reduction of peroxide is coupled to the oxidation of luminol in the presence of an enhancer molecule *(4)*. As the luminol breaks down it passes through an excited intermediate stage. Light is emitted (wavelength maximum 428 nm) as this falls to ground state, and is detected on blue-light sensitive film, producing a permanent, hard-copy result.

Fluorescein-labeled oligonucleotides may be used in a variety of applications. However their short length, compared with cloned inserts or PCR amplified material, means that their use is best suited to high target applications, such as mismatch screening by dot and slot blotting *(2)*, colony and plaque screening (*see* Fig. 1), detection of PCR products *(5)*, and DNA fingerprinting analysis (*see* Fig. 2).

2. Materials

The hybridization buffer may be stored subaliquoted at –20°C for at least 6 mo. The fluorescein-labeled oligonucleotide may be stored at –20°C, in the dark, for at least 6 mo. The antibody conjugate, BSA, and the ECL detection reagents should all be stored between 2 and 8°C. All the other reagents should be stable at room temperature. Membrane blots should be stored at room temperature in a vacuum desiccator for maximum sensitivity.

1. Blocking agent (Amersham [Amersham, UK] RPN3023).
2. Hybridization buffer component: *N*-lauryl sarcosine.
3. 20X SSC: 3.0*M* NaCl, 0.3*M* trisodium citrate, pH 7.0.
4. 10% (w/v) SDS stock solution.

Fig. 1. Colony screening. Top three filters: *E. coli* colonies transformed with pSP65 containing a portion of the human Nras proto-oncogene lifted onto Hybond-N$^+$. Bottom two filters: *E. coli* colonies transformed with pSP65 lifted onto Hybond-N$^+$. Both sets were hybridized with 10 ng/mL of an Nras specific 25-base oligonucleotide labeled at the 3' end with fluorescein-dUTP. Hybridized for 2 h at 42°C; stringent wash 1X SSC, 0.1% SDS at 42°C ; 10-min exposure on Hyperfilm-MP.

5. Hybridization buffer consisting of 5X SSC, 0.1% (w/v) hybridization buffer component, 0.02% (w/v) SDS, and 0.5% (w/v) blocking agent. Combine all the components, make up to the required volume, and, with constant stirring, heat gently until fully dissolved (about 20–30 min).

6. Fluorescein-labeled oligonucleotide probe (*see* Chapter 16).

7. Wash buffer: 5X SSC, 0.1% (w/v) SDS.

8. Stringent wash buffer: 1X SSC, 0.1% (w/v) SDS.

9. Tris buffered saline (TBS): 0.1M Tris-HCl, pH 7.5, 0.4M NaCl.

10. Membrane blocking solution: 0.5% (w/v) blocking agent in TBS. Dissolve blocking agent by gentle heating and constant stirring. Allow to cool to room temperature before use.

11. BSA Fraction V (e.g., Sigma [St. Louis, MO] A-2153, BDH [Poole, UK] Product No.4415).

12. Antifluorescein/HRP conjugate (Amersham RPN3022). Dilute this reagent, just prior to use, 1000-fold in TBS containing 0.5% (w/v) BSA.

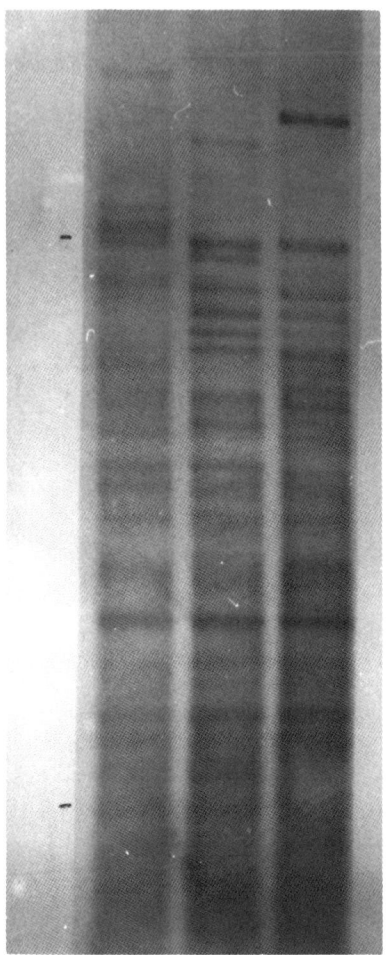

Fig. 2. Fingerprint analysis. Three separate human genomic DNA samples were digested with *Hin*fI and transfered to Hybond-N⁺ (2 µg loadings). Hybridized at 10 ng/mL with a 28-base oligonucleotide probe, specific for the insulin hypervariable repeat region, labeled at the 3' end with fluorescein-dUTP. Hybridized for 2 h at 42°C; stringent wash 1X SSC, 0.1% SDS at 42°C; 30-min exposure on Hyperfilm-MP.

13. ECL detection reagents 1 and 2 (Amersham RPN2105):
    a. Detection reagent 1 contains a peracid salt as a substrate for the HRP enzyme.
    b. Detection reagent 2 contains luminol and an optimized enhancer.

14. Nylon or nitrocellulose membrane: The system gives excellent results with Hybond-N$^+$ (nylon). Nylon membrane supplied by other manufacturers and nitrocellulose membranes may also give good results.
15. X-ray film cassette: (e.g., Hypercassettes, Amersham).
16. SaranWrap^{TM} (Dow Chemical Company)—often available from local distributors. Other cling wraps may be suitable but would have to be tested first.
17. Blue-light sensitive autoradiography film (e.g., Hyperfilm^{TM}-ECL, Amersham RPN 2103/2104).

3. Methods
3.1. Hybridization and Stringent Washes

1. Place the blot(s) into the hybridization buffer and prehybridize at the proposed hybridization temperature (*see* Note 4) for 15–30 min, in a shaking water bath. Allow 0.25 mL buffer/cm^2 membrane if boxes are used, or 0.125 mL/cm^2 if using bags (*see* Note 1).
2. Add the labeled oligonucleotide probe to the prehybridization solution to give a final concentration of 5–10 ng/mL (*see* Note 2).
3. Hybridize at the required temperature for a minimum of 2 h in a shaking water bath (*see* Note 3). For many systems a temperature of 42–50°C is suitable (*see* Note 4).
4. After hybridization place the blot(s) in a clean container and wash in an excess of wash buffer twice for 5 min at room temperature (*see* Note 5).
5. Replace the solution with an excess of stringent wash buffer (*see* Note 6). Wash at the desired temperature (typically 42–50°C) twice for 15 min in a shaking water bath.

3.2. Membrane Blocking, Antibody Incubation, and Washes

All the following steps are performed at room temperature, and all incubations require constant agitation of the blot(s).

1. Place the blot(s) in a clean container and rinse with TBS for 5 min to remove residual SDS.
2. During the stringent washes prepare the membrane blocking solution.
3. Incubate the blot(s) in membrane blocking solution for 30 min, allowing 0.25 mL/cm^2 membrane (*see* Note 7).
4. Rinse the blot(s) briefly in TBS.
5. Dilute the antifluorescein/HRP conjugate stock.
6. Incubate the blot(s) in the diluted conjugate for 30 min, allowing 0.25 mL/cm^2.

7. Wash the blot(s) four times for 5 min in an excess of TBS to remove nonspecifically bound antibody (*see* Note 8).

3.3. Signal Generation and Detection

The following steps should be executed with reasonable speed. Avoid the use of powdered gloves at this stage, since the powder may inhibit the ECL reaction and cause blank patches on the autoradiograph.

1. Mix equal volumes of the ECL detection reagents to give sufficient volume to cover the blot (approx 0.125 mL/cm^2).
2. Drain the excess wash buffer from the blot(s) and lay them DNA side up on a piece of Saran Wrap, taking care not to let them dry out (*see* Note 9).
3. Apply the mixed detection reagents directly to the blots and incubate for 1 min.
4. Drain the blot(s) on a piece of absorbent paper and wrap them in a fresh piece of Saran Wrap; try to avoid air pockets.
5. Place blots DNA side up in a film cassette and, in the dark, expose them to a sheet of autoradiography film (*see* Note 10).
6. Optimum exposure time should be found empirically and will depend on the level of target material (*see* Notes 11 and 12).

4. Notes

1. The volume of buffer during prehybridization and hybridization must be adequate to ensure free movement of the blot(s), especially if multiple blots are being hybridized together.
2. Do not apply the probe directly to the blot(s) as this may lead to high backgrounds. An alternative technique is to remove a small aliquot of the buffer, mix with the probe, then return to the container. Nonspecific background may result if the probe concentration is too high.
3. Hybridization is normally complete in 2 h, but may be extended to overnight if this is convenient.
4. The T_m of a hybrid generally depends on the salt concentration of the buffer and washing solutions, and also on the base composition. For perfectly matched oligonucleotides of between 11 and 22 nucleotides the T_m can be found from the Wallace rule (*1*):

$$T_m \, (°C) = 2°C/AT + 4°C/GC$$

A hybridization temperature of 5°C below the T_m is usually chosen for perfectly matched sequences, but for every mismatch a further 5°C decrease in temperature is necessary to maintain hybrid stability.

5. Allow 2 mL of all wash solutions/cm^2 membrane. Ensure that any precipitate (e.g., SDS) is fully dissolved by gentle warming before use, otherwise increased backgrounds may result.

6. The SSC concentration in the stringent wash can be varied from 0.1–2X SSC, while keeping the SDS concentration constant at 0.1%(w/v), to achieve the desired stringency. A temperature of 42–50°C is usually sufficient to discriminate between perfectly matched hybrids and mismatches, although it can be increased to 65°C if necessary. Nonspecific background may result if the stringency is insufficient.

7. Increasing the temperature of the blocking solution to 30–40°C prior to use may help, if high backgrounds are a problem.

8. After the antibody washes it is possible to store the blot(s) in the final wash buffer for up to 30 min prior to signal generation. Alternatively, they may be wrapped in Saran Wrap, to avoid drying out, and be stored overnight at 2–8°C.

9. Patchy background may occur if part of the membrane dries out at any stage in the procedure. Blots should only be handled with clean, blunt forceps to avoid damage and contamination.

10. During the detection stage ensure that there is no free detection fluid in the film cassette as this will cause high backgrounds. The film must be kept dry.

11. The length of exposure may be judged by doing a short, initial exposure of, for instance, 5 min and assessing the signal intensity and background. For colonies and plaques 10 min may be sufficient, whereas a fingerprint analysis may require 1 h or more.

12. If low sensitivity is a problem a number of points should be checked:
    a. Ensure adequate fixation of the DNA to the membrane.
    b. Check that the probe concentration is not too low.
    c. Ensure that the posthybridization washes are not too stringent.
    d. Check that the antibody conjugate is not overdiluted.
    e. Avoid crosscontamination of the detection reagents. These can be checked by mixing a small volume of each. In a darkroom add 5 µL of the antibody conjugate. Blue light should be visible. If not, then crosscontamination may have occurred.

References

1. Wallace, R. B., Shaffer, J., Murphy, R. F., Banner, J., Hirose, T., and Itakura, K. (1979) Hybridization of synthetic oligodeoxyribonucleotides to ØX174: the effect of single base pair mismatch. *Nucleic Acids Res.* **6,** 3545–3556.
2. Durrant, I. (1991) Stringency control with 3' tailed oligonucleotide probes. *Highlights* **2,** 6,7.

3. Thein, S. L. and Wallace, R. B. (1986) The use of synthetic oligonucleotides as specific hybridization probes in the diagnosis of genetic disorders, in *Human Genetic Diseases: A Practical Approach* (Davies, K. E., ed.), IRL, Oxford, pp. 33–50.
4. Durrant, I. (1990) Light based detection of biomolecules. *Nature* **346,** 297,298.
5. van der Auwera, B. and Gieles, M. (1991) ECL 3'oligolabeling system: HLA typing. *Highlights* **2,** 5.

CHAPTER 23

Preparation of Chromosome Spreads by Root-Tip Meristem Dissection for *In Situ* Hybridization with Biotin-Labeled Probes

Angela Karp

1. Introduction

Success with hybridization of DNA sequences *in situ* to plant chromosomes depends on many factors, but of foremost importance is the quality of the chromosome preparation. If this is not of high standard in terms of the number of metaphases achieved and the cleanliness of the background, there is really little point proceeding with the time-consuming and expensive process of the *in situ* hybridization procedure. There are several methods of making good chromosome preparations suitable for *in situ* (e.g., *see* Chapter 24). The one described here has the advantage of simplicity, requiring little in terms of equipment and materials, but it does need patience and dexterity at least in the first attempts, after which the procedure soon becomes routine.

2. Materials

1. Possible pretreatments:
   a. 8-hydroxyquinoline: 0.29 g in 1 L of distilled water, dissolved at 60°C for 6 h. Store at 18–20°C.
   b. colchicine: 0.05 g in 100 mL distilled water. Store at 4°C.

From: *Methods in Molecular Biology, Vol. 28:*
Protocols for Nucleic Acid Analysis by Nonradioactive Probes
Edited by: P. G. Isaac Copyright ©1994 Humana Press Inc., Totowa, NJ

c. α-bromonaphthalene; make up stock solution of 1 mL in 100 mL absolute alcohol. Store at room temperature. Just before use, make a dilution of 10 µL stock in 10 mL distilled water.

2. Farmer's fixative solution: 3:1 ethanol:acetic acid.
3. 45% Aqueous acetic acid.
4. Acid washed cover slips: Prepare cover slips by boiling in 0.2N HCl for a few minutes, then rinsing in distilled water. Store the cover slips in 96% ethanol. Before use, dry them with a tissue. The use of the acid-washed cover slip minimizes loss of cells by adhesion to the cover slip.
5. Rubber solution (any bicycle rubber solution will suffice, e.g., Weldtite, [code 02002], C. B. Baggs Ltd. Cricklewood, London NW2 1AL).
6. Liquid nitrogen.
7. 75% Alcohol.
8. Absolute alcohol.
9. Black storage boxes: Suitable storage boxes may be obtained from Agar Scientific, 62A, Cambridge Rd., Stansted, Essex, CM24 8DA, UK (cat nos. L4111, L4163). They consist of a black plastic container with a close-fitting lid and contain a plastic slide rack handle that will carry up to 25 slides. Tape a package of silica gel inside the lid.

3. Method

1. Collect roots from germinating seeds on moist filter paper (*see* Note 1).
2. Apply a suitable pretreatment step to arrest mitosis at metaphase (*see* Note 2). It is possible to use one of the following three pretreatments: 8-hydroxyquinoline for 3–4 h at 18–20°C, colchicine for 4–5 h at 18°C, or α-bromonaphthalene for 18 h at 4°C. The latter pretreatment works best for rye.
3. Rinse in distilled water once.
4. Fix in Farmer's solution for at least 24 h (*see* Note 3).
5. Place roots in 45% (v/v) aqueous acetic acid for 5 min.
6. Transfer to distilled water.
7. Place one root onto a clean slide (*see* Note 4). Under a dissecting microscope, remove the root cap using a clean scalpel. Cut below the meristematic portion of the root tip and discard the stalk. Transfer the tip to another clean slide and add a small drop of 45% acetic acid.
8. Hold the root tip at the base with a pair of sharp watchmaker forceps and slit vertically along the length on one side using the scalpel.
9. Keeping a grip with the forceps, use the scalpel to push out the meristematic contents through the gap made by the slit. They should "puff out in a cloud."

10. Gently circulate cells away from the remaining tip and remove the latter and any large pieces that may have broken away during the dissection.
11. Add an acid-washed cover slip and press down over filter paper very gently (it is not necessary to squash).
12. Ring the cover slip with rubber solution (e.g., "Weldtite").
13. Examine the preparation under a phase contrast microscope and retain if at least five good metaphase spreads can be seen. Mark the position of the good spreads and label the slide with a diamond pencil.
14. Remove the rubber solution with a needle, mark the corners of the cover slip with the diamond pencil and then remove the cover slip by immersing the slide in liquid nitrogen (*see* Note 5) and flicking off cover slip with a razor blade.
15. Dehydrate the slide by immersing it in 75% alcohol for 20 min followed by 20 min in absolute alcohol.
16. Store slides in black boxes with silica gel taped to the lid at –20°C.

4. Notes

1. It is important not to overwater the germinating seedlings. The filter paper should be moist but there should be no surface water. Never water just prior to collecting. Determine what root length gives maximum mitotic activity for your material. In cereals, such as rye, the roots should be taken when they are 1–2 cm in length.
2. A number of chemicals will accumulate metaphases by inhibiting spindle formation (*see* Section 3. for some possibilities). Determine which pretreatment is suitable for your material.
3. Limit fixation to a few days if possible. If the roots are too fresh it is difficult to dissect them easily because they become fragile and if they are too old the roots become brittle. About 1 wk old is ideal, depending on the species.
4. It is not necessary to use subbed slides in this procedure because the roots have not been treated with enzymes that may prevent adherence of the cells to the slide.
5. Immerse the slide by holding it with long forceps (it is also best to wear protective gloves and eye protection) and lowering it into a container with liquid nitrogen. Make sure the slide is immersed slowly, or it may crack. Keep the slide immersed until well after the liquid nitrogen has ceased to bubble. Remove the cover slip quickly before the slide warms up.

CHAPTER 24

Enzymatic Treatment of Plant Material to Spread Chromosomes for *In Situ* Hybridization

Trude Schwarzacher and Andrew R. Leitch

1. Introduction

An important factor contributing to the success of DNA:DNA *in situ* hybridization is the preparation of clean chromosome spreads *(1,2)*. The chromosomes and nuclei should be well separated and free of cytoplasm, debris, and dirt (Fig. 1). Nonspecific signal is deposited on cytoplasm and cell debris, and cytoplasm can mask the chromosomes and hinder the access of probe and detection reagents. This chapter describes two techniques for preparing chromosome spreads. The technique for preparing chromosomes by the squashing method (Section 3.2.) is modified from Schwarzacher et al. *(3)*, and the dropping method (Section 3.3.) from Ambros et al. *(1)* and Geber and Schweizer *(4)*.

In both the dropping and squashing methods described here, the digested material is treated to accumulate metaphases and then fixed. Material can be stored for several weeks prior to enzyme digestion. In the squashing method, chromosomes are spread by squashing the material in 45% acetic acid between the slide and cover slip, freezing the material to the slide, and then removing the cover slip. For the dropping method, a suspension, mostly single cells, is made from the enzyme treated material which is then resuspended in alcohol/acetic

From: *Methods in Molecular Biology, Vol. 28:*
Protocols for Nucleic Acid Analysis by Nonradioactive Probes
Edited by: P. G. Isaac Copyright ©1994 Humana Press Inc., Totowa, NJ

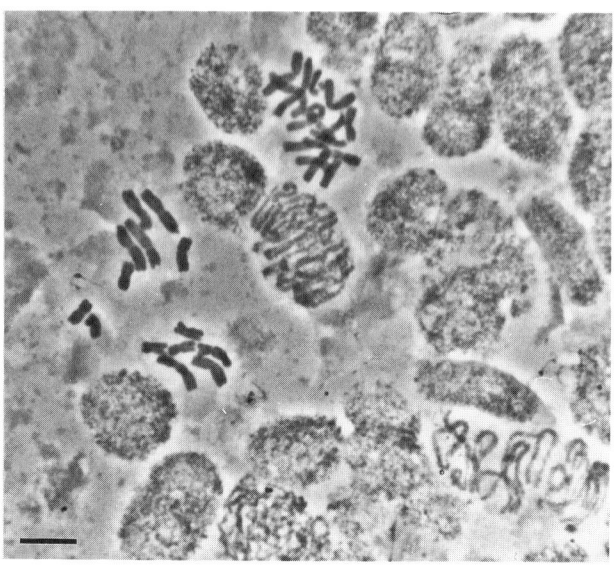

Fig. 1. Phase contrast micrograph of a squashed root tip meristem after enzyme digestion. Most of the cell walls have been degraded and the nuclei and chromosomes (dark gray to black) are liberated. Little cytoplasm and debris (lighter gray) are visible. Bar = 10 µm.

acid solution and dropped on a glass slide. As an alternative dropping method, Mouras et al. *(5)* and Murata *(6)* used living root tips or suspension cell cultures and subjected them to enzyme treatment to make a protoplast suspension that was fixed and spread. Methods of preparation of chromosomes for *in situ* hybridization not involving enzyme digestion are discussed by Karp (Chapter 23).

The dropping method has advantages over the squashing method in that the preparations are of uniform thickness and density and that several slides can be made from the same suspension batch with very similar quality. Chromosomes are more likely to spread apart free of cytoplasm and are often less distorted than in squashed preparations. The disadvantage of the dropping method is that a reasonably large quantity of material, including many dividing cells, is required. It is not possible to make a suspension for dropping from, e.g., a single root tip, in most species. In addition, species with large chromosomes or many chromosomes (e.g., hexaploid wheat, rye, or *Vicia faba*) are difficult to spread by dropping forces only. If possible we recommend using the dropping method; it might be easier to learn and we find it particularly advantageous if low or single copy *in situ* hybridization is attempted.

2. Materials

2.1. Sources of Metaphases and Materials for Fixation

1. Plant material: Any tissue containing dividing cells can be used. Root tips from young seedlings, from newly grown roots at the edge of plant pots, or hydroponic culture are all suitable. Alternatively, flower buds, anthers, carpels, or apical meristems can be used. For seed germination, put seeds in distilled water on filter paper at 20–25°C in the dark and leave until roots are about 10–20 mm long. Small seeds are best germinated under sterile conditions on agar minimal medium (e.g., Murashige and Skoog without sugar). For hydroponic growth, suspend plantlets or bulbs cleaned from soil above an aerated plant nutrient solution (e.g., Phostrogen); existing roots should be immersed in the solution, but not the plant itself. Root growth is normally initiated within a few days.
2. Metaphase arresting agents (choose one of the following; *see* Note 1):
   a. Ice water (for cereals and grasses): Put distilled water in a clean plastic bottle, shake vigorously to aerate, and keep at –20°C until the water starts to freeze, shake again.
   b. 0.05% (w/v) Colchicine (for most plant tissues).
   c. 2 m*M* 8-Hydroxyquinoline (for dicotyledonous plants, particularly those with small chromosomes, such as *Arabidopsis thaliana*).
3. Fixative: freshly prepared 3 parts 100% ethanol or methanol to 1 part glacial acetic acid.

2.2. Chromosome Preparations

1. Solution A: 0.1*M* citric acid.
2. Solution B: 0.1*M* trisodium citrate.
3. Enzyme buffer (pH 4.6): Make a 10X stock solution by mixing 40 mL solution A + 60 mL solution B. For use, dilute to 1X enzyme buffer with distilled water.
4. Enzyme solution (1X): 1% (w/v) cellulase (0.9% from *Aspergillus niger*, Calbiochem [La Jolla, CA] 21947, 0.1% Onozuka RS [Yakult Pharmaceuticals, Japan]) and 10% (v/v) pectinase (from *Aspergillus niger*, solution in glycerol, Sigma [St. Louis, MO] P-9932) in 1X enzyme buffer

(*see* Note 2). Divide the solution into 2-mL aliquots and store at –20°C. The enzyme solution can be reused several times, but should be centrifuged to remove dirt and any plant material.

5. Clean microscope slides: Place slides into chromium trioxide solution in 80% (w/v) sulfuric acid (Merck, Rahway, NJ) for at least 3 h at room temperature. Wash slides in running water for 5 min, rinse them thoroughly in distilled water, and air-dry. Place slides in 100% ethanol, remove, and dry with a tissue immediately prior to use (*see* Note 3).

6. Stereo dissecting microscope.

7. Phase contrast microscope with 10 and 40X lenses.

8. Cover slips: No.1, 18 × 18 mm (for the squashing method, *see* Note 4).

9. Fine forceps (No. 5) and dissecting needles (for the squashing method).

10. 45% Aqueous acetic acid (for the squashing method).

11. Dry ice or liquid nitrogen (for the squashing method).

12. Fixative (as described in Section 2.1. above, for the dropping method).

3. Methods
3.1. Accumulation of Metaphases and Fixation

1. Excise root tips (10–20 mm long) or other material and transfer to one of the metaphase arresting agents.

2. Incubate as follows (depending on the metaphase arresting agent; *see* Note 1): 24 h at 0°C (for ice water), or 3–6 h at room temperature or 10–20 h at 4°C (for colchicine), or 0.5–2 h at room temperature followed by 0.5–2 h at 4°C (for 8-hydroxyquinoline).

3. Briefly blot the material dry and fix immediately in fixative for at least 10 h at room temperature. If fixed material is to be kept (up to several months), leave for 2 h at room temperature and then transfer to –20°C.

3.2. Squashing Method of Chromosome Preparation

Each of the following steps, if not otherwise mentioned, is carried out at room temperature in small glass or plastic containers (5–10 mL). Material is carefully transferred by forceps or, if very small roots or buds are used, with a pipet.

1. Wash the fixed material twice for 10 min in enzyme buffer to remove fixative.

2. Transfer to 1–2X enzyme solution (1–2 mL); incubate at 37°C for 30–90 min (*see* Note 2).

3. Transfer to enzyme buffer, leave for at least 15 min.

4. Place enough material for one preparation (typically one root tip, small bud, or anther) in 45% acetic acid in an embryo dish or small Petri dish for a few minutes.

5. Make the chromosome spreads. Under the stereo microscope in 1–2 drops (10–30 µL) of 45% acetic acid on a clean slide, tease the material to fragments with a fine needle, isolate the meristem, and remove all other tissue from the slides, in particular the root cap, which is tough and prevents squashing (*see* Note 5). Apply a cover slip. Carefully tap the cover slip with a needle and then gently squash the material between glass slide and cover slip (*see* Note 6). Check the preparation under a phase contrast microscope.

6. Put the slide on dry ice for 5–10 min (preferred method) or immerse into liquid nitrogen to freeze the material, then flick off the cover slip with a razor blade, and air-dry.

7. Screen the slides under phase contrast (*see* Note 7). Small chromosomes can be stained with DAPI (Chapter 26) and analyzed under the fluorescent microscope.

8. Slides can be stored desiccated in the refrigerator or freezer for several weeks.

3.3. Dropping Method of Chromosome Preparation

After initially washing the fixed material, protoplast isolation is carried out in 1.5-mL microcentrifuge tubes at room temperature. If plenty of material is available, 10- to 15-mL polypropylene or glass centrifuge tubes can be used. If no metaphases are found, a squash preparation is required to check the material.

1. Wash the fixed material twice for 10 min in 5–10 mL of enzyme buffer to remove fixative.

2. Remove the dividing tissue, e.g., the tip of the root (2–3 mm), and place several in 1 mL of 1–2X enzyme solution.

3. Incubate at 37°C for 1–2 h (*see* Note 2). Check the material from time to time and try to disperse cells with a glass rod or pipet tip to make a suspension.

4. Centrifuge for 3 min at 800*g*, discard the supernatant, and shake up the pellet.

5. Resuspend pellet in fresh enzyme buffer and repeat step 4.

6. Repeat step 5 twice.

7. Resuspend the pellet in fresh fixative.

8. Centrifuge for 3 min at 1200*g*, discard the supernatant, and shake up the pellet.

9. Repeat steps 7 and 8 at least twice.
10. Resuspend the pellet in fixative to a final volume of 50–100 µL.
11. Drop 10–20 µL of the suspension onto a clean glass slide, shake the slide a little, and blow gently on it (*see* Note 8).
12. Leave the slide to air-dry (about 1 h).
13. Check the slides for density and spreading under a phase contrast microscope (*see* Note 7).
14. Slides can be stored desiccated in the refrigerator or –20°C freezer for several weeks.

4. Notes

1. The response to the metaphase accumulation reagents is different from species to species and has to be established by trial and error. Representative times are given in Section 3. For best results, fix material after different times of treatment, experiment with different reagents, and check the mitotic index by making chromosome preparations. It is also important not to expose seedlings, roots, and plants to chemicals and fumes, particularly fixatives (e.g., in a cold room also used for chemical storage), to use clean labware with tight lids (disposable plastic is ideal), clean forceps, and distilled water. To increase the metaphase index further, synchronization of the dividing cell population can be attempted by subjecting seedlings or plants to a period of cold temperatures (generally 24–48 h at 4°C). The plant material is returned to 20–25°C and after a suitable recovery of roughly one cell cycle period (dependent on the species; for cereals about 25–30 h) the material is then treated as in Section 3.1. Treating for too long in arresting agents results in overcondensation of the metaphase chromosomes, which might be desirable for counting chromosomes, but not for *in situ* hybridization where spatial resolution along chromosomes is wanted.

2. The enzyme digestion step needs to be carefully adjusted to the material used, by changing the time of digestion, the strength of enzyme, and sometimes the ratio of cellulase to pectinase. Ideally, the cell walls should be weakened, such that the cells can be separated easily, and chromosome spreads are clean of cytoplasm. In most cases the meristematic cells will be digested faster than the nondividing tissue. If the squashing method is used, material should remain intact to handle, otherwise the dividing cells are lost into the medium. For the dropping method, enzyme digestion should be longer or at a higher concentration. If material has been fixed for several weeks, the material becomes harder and needs longer digestion. Onozuka cellulase is very pure and digests cellulose efficiently, so shorter digestion times should be used.

3. Slides need to be very clean and grease-free to ensure adhesion of the chromosomes and nuclei to the slides, which will have to pass through many steps during *in situ* hybridization.

4. The cover slips should be free of dirt, but should not be cleaned with alcohol or chromic acid; otherwise material will stick to the cover slip and will be lost when the cover slip is removed (step 6 of Section 3.2.).

5. It is important to make a monolayer of cells and nuclei and to use the correct amount of fluid between the cover slip and glass slide for squashing. Excess fluid will move all cells to the edge of the cover slip, too little fluid will encourage air bubbles. Thick and too dense material will not spread.

6. Tapping with the needle on top of the cover slip disperses cells, but care needs to be taken not to shear the material by lateral movement of the cover slip. The final thumb pressure to squash the material should be moderately strong, but not abrupt.

7. It is advisable to be very rigorous in selecting good slides. We routinely discard almost half of our spreads from good batches of fixations, and half of all batches of fixations.

8. When the fixative cell suspension is dropped onto the surface of the glass slide in the dropping method, the fixative disperses and evaporates quickly, and the cells, nuclei, and chromosomes are spread by the forces generated. This step can be monitored under phase contrast and adjusted to give well-separated chromosomes in complete metaphase plates. Shaking and blowing the slide increases the spreading forces, whereas gentle dropping and handling keep the chromosomes together. The density of the cells is also important for effective spreading. Cells should be abundant, but not overlapping or too close to each other. Repeated centrifugation and resuspension in fixative can help to decrease the amount of dirt in the suspension. If necessary, the preparation of the suspension has to be repeated with modified enzyme digestion time.

Acknowledgments

We would like to thank J. S. Heslop-Harrison and I. J. Leitch for helpful discussions and reading the manuscript, M. Shi for technical assistance, and BP and Venture Research International for support.

References

1. Ambros, P. F., Matzke, M. A., and Matzke, A. J. M. (1986) Detection of a 17 kb unique sequence (T-DNA) in plant chromosomes by *in situ* hybridization. *Chromosoma* **94,** 11–18.
2. Schwarzacher, T., Leitch, A. R., Bennett, M. D., and Heslop-Harrison, J. S. (1989) *In situ* localization of parental genomes in a wide hybrid. *Ann. Bot.* **64,** 315–324.

3. Schwarzacher, T., Ambros, P., and Schweizer, D. (1980) Application of Giemsa banding to orchid karyotype analysis. *Plant Syst. Evol.* **134,** 293–297.

4. Geber, G. and Schweizer, D. (1988) Cytochemical heterochromatin differentiation in *Sinapis alba* (Cruciferae) using a simple air-drying technique for producing chromosome spreads. *Plant Syst. Evol.* **158,** 97–106.

5. Mouras, A., Saul, M. W., Essad, S., and Potrykus, I. (1987) Localization by *in situ* hybridization of a low copy chimaeric resistance gene introduced into plants by direct gene transfer. *Mol. Gen. Genet.* **207,** 204–209.

6. Murata, M. (1983) Staining air dried protoplasts for study of plant chromosomes. *Stain Technology* **58,** 101–106.

CHAPTER 25

Use of Biotin-Labeled Probes on Plant Chromosomes

Angela Karp

1. Introduction

In situ hybridization of biotin-labeled probes to plant chromosomes is a powerful technique enabling the physical mapping of DNA sequences and the "tagging" of chromosomes for identification *(1)*. The method described here is based on Rayburn and Gill *(2)* and is particularly successful for detection of sequences present in multiple copies. In this method the signal is detected using a streptavidin/horseradish peroxidase conjugate, which binds to the biotin-labeled probe, and the color reaction is produced by the addition of hydrogen peroxide and the dye, 3,3'-diaminobenzidine tetrahydrochloride dihyhdrate (DAB). The signal is seen as a dark precipitate (Fig. 1) that maintains color intensity for several years, provided slides are kept away from bright light. Alternative methods of probing and detection are given in Chapters 26 and 27. The whole procedure can be run in 1 d, with a 4-h hybridization period, but we prefer to begin the method in the afternoon and to hybridize overnight.

2. Materials

1. All glassware and Eppendorf tubes should be silanized (*see also* Chapter 13). For Eppendorf tubes, fill a glass beaker containing Eppendorf microcentrifuge tubes with "Repelcote," a silicone water-repellent from BDH (Poole, UK) (N.B.: This material is harmful and the procedure should

From: *Methods in Molecular Biology, Vol. 28:*
Protocols for Nucleic Acid Analysis by Nonradioactive Probes
Edited by: P. G. Isaac Copyright ©1994 Humana Press Inc., Totowa, NJ

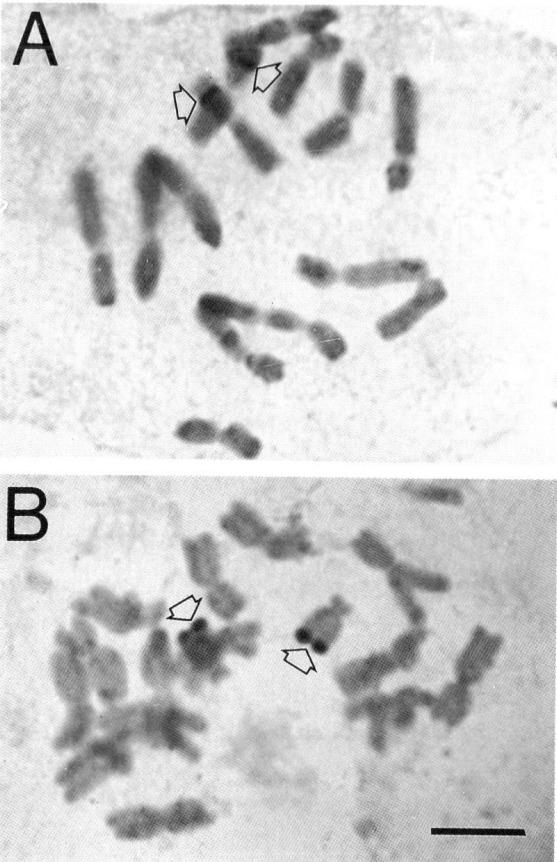

Fig 1. *In situ* hybridization using biotin-labeled probes. Signal detection was achieved using a streptavidin-horseradish peroxidase conjugate, which binds to the biotin followed by addition of hydrogen peroxide and the dye 3,3'-diaminobenzidine tetrahydrochloride dihydrate (DAB). **(A)** Localization of ribosomal RNA genes in rye. Two sites are detected (arrows) corresponding with the nucleolar organizers on chromosome 1 (A. Karp, unpublished results). **(B)** Localization of a B-chromosome specific probe (arrows) to the two B chromosomes of a rye plant *(3)*. (Scale bar = 10 µm).

be carried out in a fume-hood). Swirl round and discard it and repeat the treatment. Wash the Eppendorf tubes several times with water before autoclaving and drying in an oven at 80°C. For glassware, carry out the same procedure except, after the treatment, air-dry and sterilize in an oven at 180°C.

2. RNaseA: Prepare a 10 mg/mL stock solution in sterile distilled water. Heat to 100°C for 15 min in a water bath. Store at –20°C. Dilute the stock solution of RNaseA (10 mg/mL) to 200 µg/mL (i.e., 1/50) using 2X SSC with 1% bovine serum albumin (BSA). Store in 1-mL aliquots at –20°C.

3. 20X SSC: 100.2 g trisodium citrate, 175.0 g sodium chloride, 1.0 L double distilled water. Autoclave.

4. Dextran sulfate: 500 mg/mL aqueous solution. Take 0.5 mL of sterile distilled water, add to 500 mg dextran sulfate in a microcentrifuge tube. Mix and add the remaining 0.5 mL sterile distilled water. Vortex on a whirlimixer. Boil in a water bath. Repeat the last two steps until the dextran sulfate is dissolved.

5. Herring sperm carrier DNA: 10 mg/mL in distilled water. Mix well by stirring for 5–10 min. Store in 1-mL aliquots in Eppendorf tubes at –20°C.

6. Deionized formamide: 200 mL formamide, 1 g Dowex 1 (chloride form) (base anion), 1 g Dowex 50W (hydrogen form). Stir for 1 h. Filter through Whatman filter paper no. 1.

7. 70% Formamide: 70 mL deionized formamide, 30 mL 2X SSC

8. Phosphate buffered saline (PBS): 200 g NaCl, 5 g KCl, 5 g KH_2PO_4, 37 g $Na_2HPO_4 \cdot 12 \ H_2O$, 1.0 L double distilled water. Autoclave.

9. BRL Streptavidin/horseradish peroxidase conjugate (Cat. no. 9534SA). Add 8 µL of the conjugate to 792 µL 0.05M Tris-HCl, pH 7.6. Prepare this immediately before use (Section 3., step 10).

10. 0.05M Tris-HCl, pH 7.6, 0.01M imidazole.

11. DAB = 3,3'-diaminobenzidine tetrahydrochloride dihydrate, Cat. no. 5972SA from BRL. **Caution:** Handle with extreme care in fume-hood using protection for eyes and hands. For a stock solution, add 10 mL of 0.05M Tris-HCl buffer, pH 7.6, to 100 mg DAB powder. Mix well. Transfer in 1-mL aliquots to Eppendorf tubes. Wrap each Eppendorf tube in foil (DAB is light sensitive) and store at –20°C. A 0.05% solution should be made up just before use: Mix 575 µL of 0.05M Tris-HCl, pH 7.6, 0.01M imidazole, plus 33.75 µL DAB stock solution and 18.75 µL hydrogen peroxide.

12. Giemsa buffer tablets, pH 6.8, from British Drug Houses, Broom Rd., Parkstone, Poole, BH12 4NN, UK (BDH) Cat. no. 33199.

13. Leishman's solution: 1 g Leishman's powder, 500 mL methanol. Mix and incubate overnight at 37°C. Store in the dark. Use as 1:3 dilution with phosphate buffer, pH 6.8, made using buffer tablets.

14. 2X SSC: Dilute stock solution of 20X SSC; 100.2 g trisodium citrate, 175.0 g sodium chloride, 1.0 L double distilled water.

15. 70% Ethanol.

16. 96% Ethanol.

17. Absolute alcohol.

18. Acid washed cover slips: Prepare cover slips by boiling in 0.2*N* HCl for a few minutes, then rinsing in distilled water. Store the cover slips in 96% ethanol. Before use, dry them with a tissue.

19. Rubber solution: "Weldtite" code no. 02002, C. B. Baggs Ltd. Cricklewood, London NW2 1AL.

20. Euparol.

21. Biotin-labeled probe (*see* Chapter 13): The probe should be in 20 µL of deionized formamide. If it has been stored in TE buffer, ethanol precipitate it, dry it under vacuum, and redissolve in 20 µL of deionized formamide.

22. Chromosome spreads (*see* Chapters 23 and 24).

3. Method

Wear gloves throughout the procedure for safety reasons.

1. Remove slides from freezer store (*see* Note 1) and lay on paper towels until dry. Label each slide with the probe details.

2. Add 100 µL RNaseA solution (200 µg/mL in 2X SSC+ 1% BSA) to each preparation. Cover with a 22 × 22 mm cover slip.

3. Incubate slides at 37°C for 30 min in a moist chamber (*see* Note 2).

4. After the RNaseA treatment, wash the slides in the following series 2X SSC, 2X SSC, 70% ethanol, and 96% ethanol for 5 min each at room temperature (*see* Note 3). The cover slip will float off during the first wash. Lay the slides on paper towels to air-dry for 30 min.

5. While the slides are drying, denature the biotinlyated probe. Heat up a beaker of water to boiling. To the probe (in 20 µL of deionized formamide in a microcentrifuge tube) add the following (*see* Note 4): 20 µL deionized formamide, 8 µL 20X SSC, 16 µL dextran sulfate (*see* Note 5), 4 µL carrier DNA, 12 µL sterile distilled water. Centrifuge the probe cocktail for few seconds. Tape the lid of the tube down to secure it and place the tube in a float. Drop the float into the beaker of boiling water for 5 min. Remove the tube and quench immediately on ice. Keep on ice until required.

6. After air-drying the slide for 30 min, denature the slides by immersing them, one at a time, in 70% formamide at 70°C for exactly 2.5 min (*see* Note 6). Immediately dehydrate each slide by passing through a series of 70%, 96%, and 100% alcohol for 5 min each on ice.

7. As soon as the alcohol has dried, immediately add 9.5 µL of probe cocktail to each slide (*see* Note 5). Add an acid-washed 22 × 22 mm cover slip, trying to avoid air bubbles, and ring with rubber solution.

8. Place the slides on a float and incubate at 37°C for 4–18 h (*see* Note 7) in a moist chamber (*see* Note 2).

9. After the hybridization, remove the rubber solution with a needle and float off the cover slips in 2X SSC. Place the slides in a rack (*see* Note 3)

and pass through following series of washes: 2X SSC for 5 min at room temperature, 2X SSC for 10 min at 37°C, 2X SSC for 5 min at room temperature, PBS for 5 min at room temperature.

10. During the last wash, prepare the steptavidin/peroxidase solution (*see* Section 2., item 9)

11. After the PBS wash, shake the surface moisture from the slides and add 100 μL of the conjugate to each slide. Cover with 22 × 22 mm cover slip, avoiding air bubbles, and incubate at 37°C for 30 min.

12. Wash the slides in PBS for 5 min at room temperature, and then air-dry at room temperature. Immediately after the slides have dried (*see* Note 8), add 78 μL of a 0.05% DAB solution. Cover with a 22 × 22 mm unwashed cover slip and incubate at 37°C for 5 min in a foil-covered moist chamber.

13. Dip the slides in PBS to remove the cover slip and rinse briefly in PBS.

14. Stain one slide at a time in Leishman's for 5 s only. Rinse briefly in buffer and examine under a light microscope. Stain longer if necessary.

15. Air-dry overnight and the following day mount in "Euparol" by adding a drop of Euparol onto the slide and covering with a cover slip.

4. Notes

1. A method of making chromosome preparations for *in situ* hybridization by dissection of root-tip meristems is given in Chapter 23. The prepared slides are stored in plastic slide boxes, with a silica gel parcel taped to the lid, at –20°C for up to 2 wk. The boxes are obtained from Agar Scientific, 62A, Cambridge Rd., Stansted, Essex, CM24 8DA, UK (cat. nos. L4111, L4163). It is not advisable to use slides that have been stored for less than 1 wk or for longer than 6 wk.

2. For a moist chamber we use a square Tupperware container with a close-fitting lid. Water is placed at the bottom of the container to a depth of 1 cm and the slides are laid flat on plastic platforms raised above the water level.

3. The black plastic slide boxes, which are supplied complete with slide racks, are used for all the washes in the *in situ* hybridization procedure. The slides can be slotted into a rack and moved in the rack from wash to wash.

4. The probe cocktail is sufficient for eight slides. It is not advisable to handle more than eight slides in one experiment but two different probes can be tested on four slides each, in which case, the amounts added to each probe should be halved.

5. Dextran sulfate is very viscous and it is necessary to cut the end of the pipet tip in order to transfer accurately.

6. The denaturing time given here is suitable for rye, maize, wheat, barley, potato, and *Brassicas*. The time needed may vary depending on the plant species from 2–3 min.

7. The minimum time required for hybridization is 4 h. Overnight is preferable in our opinion.

8. It is imperative that the slides are not allowed to dry out completely from this step onwards, otherwise there will be unspecific background labeling.

References

1. Hutchinson, J. and Lonsdale, L. M. (1982) The chromosomal distribution of cloned highly repetitive sequences from hexaploid wheat. *Heredity* **48,** 371–376.
2. Rayburn, A. L. and Gill, B. S. (1985) Use of biotin-labeled probes to map specific DNA sequences on wheat chromosomes. *J. Heredity* **76,** 78–81.
3. Blunden, R., Wilkes, T. J., Forster, J. W., Sanders, M. J., Karp, A., and Jones, R. N. (1993) Identification of the E3900 family, a second family of rye B-chromosome specific sequences. *Genome* (in press).

CHAPTER 26

Direct Fluorochrome-Labeled DNA Probes for Direct Fluorescent *In Situ* Hybridization to Chromosomes

Trude Schwarzacher and J. S. (Pat) Heslop-Harrison

1. Introduction

Chromosomal *in situ* hybridization enables determination of the presence and location of DNA sequences complementary to a labeled probe along chromosomes and within interphase nuclei. The use of directly fluorochrome-labeled probes means that sites of probe hybridization can be directly visualized. Lengthy detection procedures, as in the indirect immunochemistry methods to detect biotin or digoxigenin, are not needed. However, because there is no signal amplification, sensitivity is limited and the system is used mostly to detect middle- to high-copy number (Fig. 1) or pooled probes, although antibody amplification of some fluorochrome labels is possible *(1)*, and then the method must be evaluated in comparison with digoxigenin (Chapter 27) or biotin (Chapter 25).

Until recently, the application of direct fluorescent labeling *in situ* hybridization methods was limited, perhaps because the low sensitivity was overemphasized and commercial labeling systems were not available (complex chemistry was required to label the probe or nucleotides). The combination of enzymic labeling techniques and commercial fluorochromized nucleotides, with a better understanding

From: *Methods in Molecular Biology, Vol. 28:*
Protocols for Nucleic Acid Analysis by Nonradioactive Probes
Edited by: P. G. Isaac Copyright ©1994 Humana Press Inc., Totowa, NJ

of conditions for *in situ* hybridization and improved imaging methods, have now boosted direct fluorochrome-labeling methods as an alternative to the well-established indirect methods *(2,3)*. It is worth noting that one of the first nonradioactive *in situ* hybridization procedures used directly labeled probes *(4)*.

Advantages of the direct fluorochrome-labeling method include simplicity, as no detection steps are necessary after posthybridization washes, and less background noise as immunocytochemistry is not needed (Fig. 1). Signal quantification is more reliable since the stoichiometry of the single hybridization step is clearer than that of a multistep immunocytochemical procedure. Multiple probes, each labeled with a different fluorochrome combination, allow simultaneous multitarget detection. Three labels are currently available allowing easy multidetection of several probes (Fig. 1). In addition fluorochrome-labeled DNA can be hybridized together with digoxigenin- or biotin-labeled probes, increasing the number of simultaneously detected probes *(3)*. The antibody detection steps necessary for biotin or digoxigenin do not reduce the fluorescence of the fluorochrome-labeled probes. We now routinely use the combination of a rhodamine-labeled probe with FITC detection of a digoxigenin-labeled probe.

Disadvantages of the direct fluorescent *in situ* hybridization include lower sensitivity than with alternative methods (digoxigenin or biotin); although low-light imaging cameras and antibody-amplification can be used to overcome this problem. *In situ* hybridization signal can only be detected using a good epi-fluorescent microscope and a sensitive imaging system (film or video camera) might be required. Direct fluorochrome-labeled probes cannot be used for *in situ* hybridization at the electron microscopic level without secondary detection.

The direct fluorescent *in situ* hybridization procedure follows general *in situ* hybridization protocols (e.g., *5*) and involves labeling of the probe (*see* Chapter 15) and making the chromosome preparation (*see* Chapters 23 and 24). The steps described in this chapter cover pretreatment of chromosome spreads to reduce nonspecific probe and detection binding, increase probe permeability, and stabilize the target DNA sequences (Section 2.1. and 3.1.). Then the hybridization mixture, including the probe, blocking DNAs (if desired), formamide, and salts is prepared (Section 2.2.). Denaturation of probe and chromosomes to make them single stranded enabling hybridization of the probe to

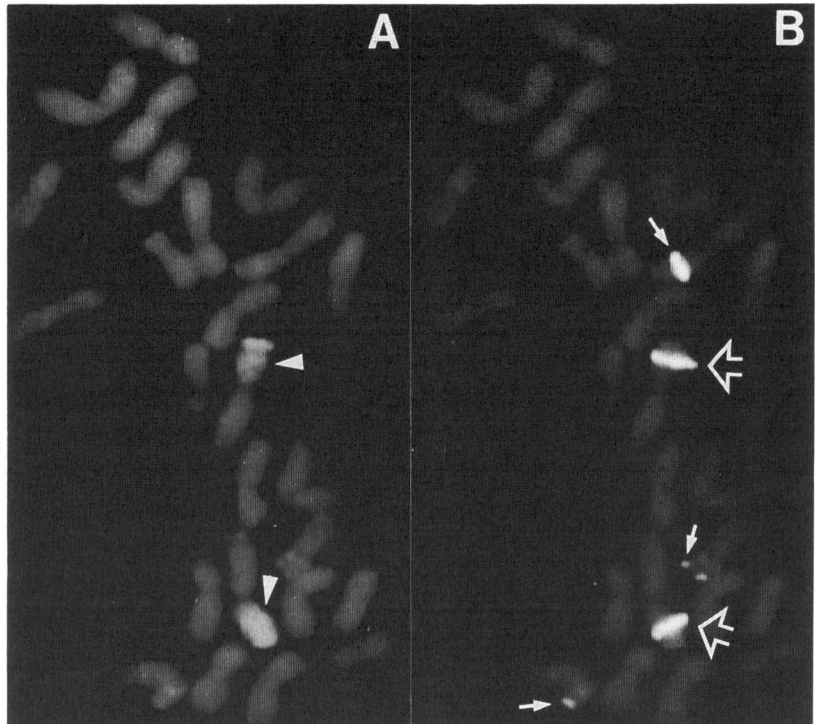

Fig. 1. Direct fluorescent double *in situ* hybridization to a partial metaphase of a hexaploid wheat variety carrying a rye translocation (1B/1R). **(A)** Bright fluorescence of the rye chromosome arms probed with rhodamine-labeled total genomic DNA from rye after green excitation (arrow heads), with the wheat chromosomes fluorescing weakly. **(B)** Strong fluorescence of the fluorescein-labeled rDNA under blue excitation. rDNA sites can be distinguished on several wheat chromosomes (arrows) and the two rye arms (open arrows). Bar = 10 μm.

the chromosomes *in situ* is described in Section 3.2. After hybridization, the weakly bound probe is removed in a series of wash steps (Section 3.2.). The stringent wash allows sequences with more than about 85% homology to remain hybridized, but can be altered if desirable. Finally, counterstaining of the chromosomes and microscopic examination is described (Section 3.3.).

Table 1 lists the different fluorochromes used as conjugates to nucleotides or antibodies, and as DNA stains for chromosomes. Fluorochromes are molecules that are excited by light at a certain wavelength and then emit light of longer wavelength (of lower energy). The most common types have characteristics of UV excitation—blue fluorescence, blue excitation— yellow/green fluorescence, and green excitation—red fluorescence. Epi-fluorescence microscopes use lamps and filter blocks to select light of the correct wavelength for excitation and emission. If several fluorochromes (Fig. 1) are used simultaneously, each needs to have a different pair of excitation and/or emission maxima.

2. Materials

Unless indicated, all solutions should be made up freshly in deionized or distilled water. All procedures and washes are at room temperature unless otherwise stated. Good laboratory practice is suitable for handling all dilute solutions; paraformaldehyde fixation and formamide washes should be carried out in a fume-hood.

2.1. Chromosome Spread Pretreatment

1. Plant chromosome preparations: Follow the methods in Chapters 23 or 24.
2. 20X SSC: 3M NaCl, 0.3M Na citrate adjusted to pH 7 with HCl and stored at room temperature.
3. 2X SSC: Dilute 20X SSC 1:10 with distilled water.
4. RNase A: Prepare a stock solution by dissolving 10 mg/mL of DNase-free RNase in 10 mM Tris-HCl, pH 7.5 and 15 mM NaCl. Boil for 15 min and allow to cool. Store at –20°C in aliquots. For use dilute the stock 1:100 in 2X SSC to give a final concentration of 0.1 mg/mL (*see* Note 1).
5. 0.01M HCl.
6. Pepsin (porcine stomach mucosa, activity: 3200 to 4500 U/mg protein, Sigma, St. Louis, MO): Prepare a stock solution by dissolving 500 µg/mL in 0.01M HCl. Store at –20°C in aliquots. For use dilute stock 1:100 in 0.01M HCl (to give a final concentration of 5 µg/mL) (*see* Note 2).
7. Depolymerized paraformaldehyde is prepared in a fume-hood by adding 2 g of paraformaldehyde to 40 mL water, heating to 60°C, adding up to 10 mL 0.1M NaOH until the solution clears, and adjusting the final volume to 50 mL with water. The paraformaldehyde solution must be freshly prepared each time (*see* Note 3).
8. 70, 90, and 100% Ethanol.

2.2. Denaturation, Hybridization, and Posthybridization Washes

1. Probe: Label the probe DNA with direct fluorochrome-labeled nucle-otides (available from Amersham [Amersham, UK] or Boehringer Mannheim [Mannheim, Germany]) as described in Chapter 15. Incor-

Table 1
Properties of Fluorochromes Used in Signal Generating Systems for *In Situ* Hybridization and as Stains for Chromosomal DNA

Fluorochrome	Excitation (max. nm)	Emission (max. nm)	Fluorescence color
Conjugated directly to nucleotides			
Coumarin (AMCA)[a]	350	450	Blue
Fluorescein (FITC)[b]	495	515	Green
Rhodamine600 (TRITC)[c]	575	600	Red
DNA stains			
DAPI[d]	355	450	Blue
Hoechst 33258	356	465	Blue
Chromomycin A3	430	570	Yellow
Quinacrine	455	495	Yellow-green
Propidium iodide	340,530	615	Red

[a]7-amino-4-methyl-coumarin-3-acetic acid; [b]Fluorescein isothiocyanate; [c]Tetramethyl rhodamine isothiocyanate; [d]4',6-diamidino-2-phenylindole.

poration of fluorochrome labels into the probe should be tested by putting a small drop of probe on a microscope slide and observing it under the epi-fluorescence microscope using the correct filter block (Table 1). The whole drop should fluoresce uniformly in the appropriate color.

2. Salmon sperm DNA (5 µg/µL): Autoclave to fragments of 100–300 bp (typically 5 min at 100 kgm^{-2}) and store in aliquots at –20°C.

3. 20X SSC (*see* Section 2.1.): Filter (0.22 µm) sterilize about 5 mL and store at –20°C, for use in the hybridization mix.

4. Formamide: Good but not the highest grade (e.g., Sigma catalog no. F 7508). Store in aliquots at –20°C.

5. 50% (w/v) Dextran sulfate in water: This takes some time to dissolve and can be speeded up by heating to 70°C. The solution is very viscous and filter (0.22 µm) sterilizing needs considerable hand pressure. Store in aliquots at –20°C.

6. 10% (w/v) Sodium dodecyl sulfate (SDS) in distilled water: Filter (0.22 µm) sterilize.

7. Hybridization mixture is prepared (*see* Note 4) by adding 25–100 ng probe DNA (*see* Note 5), 1–5 µg salmon sperm DNA (*see* Note 6), 20 µL formamide (final concentration 50%), 4 µL 20X SSC (final concentration 2X, *see* Note 7), 8 µL dextran sulfate (final concentration 10%), and 0.5 µL SDS (final concentration 0.1%). Make up with distilled water to the

final volume of 40 µL that is required per slide.

8. Humid chamber: Place filter paper in the bottom of a metal tin with a lid and soak with 2X SSC or distilled water.

9. Plastic cover slips: pieces of appropriate size cut from polyethylene autoclavable waste disposal bags.

10. 2X SSC (*see* Section 2.1.). Preheat to 42°C.

11. Stringent wash: 20% (v/v) formamide in 0.1X SSC. Preheat to 42°C (*see* Note 7).

2.3. Counterstaining and Mounting of In Situ *Hybridization Preparations*

1. McIlvaine's buffer (pH 7.0): Solution A: $0.1M$ citric acid. Solution B: $0.2M$ Na_2HPO_4. Mix 18 mL of A and 82 mL of B. Store in aliquots at –20°C.

2. Phosphate buffered saline (PBS): $0.13M$ NaCl, $0.007M$ Na_2HPO_4, $0.003M$ NaH_2PO_4, adjust to pH 7.4.

3. DAPI (4',6-diamidino-2-phenylindole): Prepare a stock solution of 100 µg/mL in water. DAPI is a potential carcinogen. To avoid weighing out the powder, order small quantities and use the whole vial to make the stock solution. Aliquot and store at –20°C (it is stable for years). Prepare a working solution of 2 µg/mL by dilution in McIlvaine's buffer, aliquot, and store at –20°C (*see* Note 8).

4. Propidium iodide (PI): Prepare a stock solution of 100 µg/mL in water. PI is a potential carcinogen. To avoid weighing out the powder, order small quantities and use the whole vial to make the stock solution. Aliquot 50 or 100 µL in 1.5-mL microcentrifuge tubes and store at –20°C. Dilute with 2X SSC or PBS to 2–5 µg/mL prior to use. PI does not keep in diluted form.

5. Antifade: 100 mg p-phenylenediamine dihydrochloride (Sigma P1519) in 10 mL PBS. Adjust to pH 8 with $0.5M$ carbonate-bicarbonate buffer (0.42 g $NaHCO_3$ in 10 mL water, pH to 9 with NaOH). Mix 1:9 with glycerol for fluorescence microscopy (Merck, Darmstadt, Germany), filter (0.22 µm) sterilize, aliquot, and store at –20°C. The solution darkens with time, but still remains effective. Antifades are also available commercially: solution AF1 from Citifluor Ltd, Connaught Building, City University, Northampton Square, London EC1V OHB, UK, or from Vector Laboratories, Burlingame, CA.

6. Glass cover slips: No. 0, 24 × 40 mm or 24 × 50 mm.

7. Fluorescent photomicroscope with filter blocks for UV, blue and green excitation, fluorescence objectives 25, 40, and 100X oil. Standard procedures are required to avoid personal exposure to high power UV light and to reduce risks from bulb breakage.

3. Methods

For steps in the schedule using volumes of 200 µL or less the solution is pipeted onto the slide and covered with a plastic cover slip. All washing steps are carried out in Coplin jars using a sufficient volume of liquid to fully immerse the whole slide. Slides must be handled carefully to avoid scratching. They should never dry out during incubation or between steps (except after step 9 of Section 3.1.).

3.1. Chromosome Spread Pretreatment

Steps are carried out at room temperature unless otherwise indicated.

1. Dry the chromosome spreads in an incubator at 37°C overnight.
2. Add 200 µL of RNase A, cover, and incubate for 1 h at 37°C in a humid chamber (*see* Note 1).
3. Wash the slides in 2X SSC three times for 5 min each.
4. Place slides in 0.01M HCl for 2 min.
5. Add 200 µL of pepsin, cover, and incubate for 10 min at 37°C (*see* Note 2).
6. Stop the pepsin reaction by placing the slides in water for 2 min and then wash in 2X SSC twice for 5 min each.
7. Place slides into freshly depolymerized paraformaldehyde and incubate for 10 min (*see* Note 3).
8. Wash slides in 2X SSC three times for 5 min each.
9. Dehydrate slides 3 min each in 70, 90, and 100% ethanol and then air-dry.

3.2. Denaturation, Hybridization, and Posthybridization Washes

1. Prepare the humid chamber and place in a water bath to raise the internal temperature to 90°C (preferably monitor the temperature using a digital thermometer with the probe in the chamber). Leave to equilibrate for at least 15 min.
2. Denature the hybridization mixture at 70°C for 10 min and then transfer to ice for 5 min.
3. Add 40 µL denatured hybridization mixture to each slide and cover with a plastic cover slip. Ensure no bubbles are trapped. Quickly place the slides in the preheated humid chamber and incubate for 10 min at 90°C (*see* Note 8).
4. Transfer the humid chamber to a 37°C incubator or water bath (*see* Note 9) and leave slides to hybridize overnight (*see* Note 10).

5. After hybridization carefully remove the cover slip and place the slides in a Coplin jar containing 2X SSC at 42°C.
6. Pour off 2X SSC and replace with the stringent wash solution at 42°C (*see* Note 7). Incubate for 10 min shaking gently. Replace the stringent wash solution once during incubation.
7. Wash the slides in 2X SSC three times for 3 min at 42°C, then three times for 3 min at room temperature.

3.3. Counterstaining, Mounting, and Examination

All steps are carried out at room temperature. Depending on the fluorochromes attached to the probe (or the detection reagents), select appropriate counterstains (*see* Table 1). They enable easy screening and identification of chromosomes and nuclei. Several DNA stains can also be applied together to produce specific bands *(6)*. Use DAPI for green (fluorescein) and red (rhodamine) fluorescing probes, but not for blue (coumarin) fluorescing probes. Fluorescein detected *in situ* hybridization signal can be counterstained with propidium iodide (PI), which fluoresces red under many excitation wavelengths; the fluorescein signal will appear yellow as a result of overlapping red and green fluorescence.

1. Add either 100 µL of DAPI or 100 µL of PI per slide (*see* Notes 11 and 12), cover with a plastic cover slip and incubate for 10 min. Wash briefly in 2X SSC, and drain.
2. Apply 1–2 drops of antifade (*see* Note 13) on the wet slide, cover with a glass cover slip. Firmly squeeze excess antifade from the slide with filter paper.
3. Slides can be viewed immediately, but the signal stabilizes after storing for a few days in the dark at 4°C (*see* Note 14). Slides can be kept at 4°C for up to 1 yr.
4. Photograph areas of interest (*see* Notes 15–17).

4. Notes

1. RNase A serves to remove endogenous RNA that might bind the probe leading to a high background signal.
2. Treatment of slides with pepsin can increase probe reagent accessibility by digesting proteins. It is particularly effective if cytoplasm is associated with the chromosome preparations. Time and concentration of the pepsin treatment might need to be adjusted for different material. If the chromosome preparations are very clean, it can be left out (steps 4–6 of Section 3.1.).

3. The paraformaldehyde fixation step minimizes the loss of DNA from chromosomes during the subsequent denaturation steps.
4. The hybridization mixture is usually prepared immediately prior to use but can be stored at –20°C for up to 6 mo.
5. Increasing the probe concentration tends to result in higher levels of background signal rather than stronger *in situ* hybridization signal.
6. Blocking DNA is used at concentrations of 2–100X probe concentration. For highly repeated probes we typically use a concentration of 50X the probe concentration.
7. The stringency of the hybridization can be varied by changing the temperature and the formamide and salt concentration. Under the conditions described here (50% formamide in 2X SSC at 37°C), for labeled probes about 200–300 bp long (usual length after nick translation or random primed labeling), 75–80% homology is required between the probe and target DNA to form stable hybrids. The stringent wash (step 6, Section 3.2.) is about 5% more stringent. If stringency requirements are different, e.g., for short oligonucleotide probes *(7)*, the conditions for both the hybridization and the stringent wash should be altered accordingly.
8. An alternative method for denaturation using a modified programmable temperature controller was described by Heslop-Harrison et al. *(5)*.
9. At the end of the denaturation step the humid chamber should be transferred as quickly as possible to the incubator at 37°C so that the material does not cool down too rapidly. This will allow the most similar sequences to reanneal first.
10. Ensure that the tin contains well soaked tissues prior to leaving it for hybridization overnight otherwise the slides will dry out.
11. Alternative widely used DNA stains are discussed by Schweizer *(6)* and listed in Table 1.
12. Avoid overstaining with counterstain, since PI can obscure weak hybridization sites. If PI is too strong, carefully remove cover slip incubate slides for 1–3 min in water, then 1 min in 2X SSC, and mount again in antifade. If counterstains (DAPI or P I) are too weak, restaining is sometimes possible.
13. To prevent fading (particularly under UV excitation) use antifades, store slides in the refrigerator for a few days. Do not look too long.
14. Slides that have been stored for a few months might have high background fluorescence. They can be revitalized by carefully removing the cover slip and remounting in antifade.
15. Color print films are easier to use, since they have more exposure latitude, adjustment of contrast, and ease of use during printing, and reproduction as plates or slides is excellent. We recommend Fujicolor 400

and Kodak Ektar 1000 films. Other suitable films include Kodak TMAX 400 for black and white photography, and Agfa or Fuji slide films.

16. When taking photographs of fluorescent images that have a dark to black background and very often only fill part of the frame, it is necessary to adjust the exposure time. Set the automatic exposure unit to dark field or equivalent, if possible, or rate the film at least two stops more sensitive than it is. It is advisable to bracket the exposure time (it should range between a few seconds and half a minute) and to make a test film. The different fluorescent colors and the various films will need different corrections.

17. Fluorescent *in situ* hybridization signals fade easily (*see also* Note 13) and normally only one or two exposures can be taken from each image. Photograph hybridization signal first and then DNA counterstain; photograph excitation with the longest wavelength first, and UV last.

Acknowledgments

We thank I. J. Leitch and A. R. Leitch for helpful discussion. We thank M. Shi and G. E. Harrison for technical assistance and BP Venture Research for support.

References

1. Bauman, J. G. J., Wiegant, J., and Van Dujin, P. (1981) Cytochemical hybridization of fluorochrome labeled RNA. III. Increased sensitivity by the use of anti-fluorescein antibodies. *Histochemistry* **73**, 181–193.

2. Wiegant, J., Ried, T., Nederlof, P. M., Van der Ploeg, M., Tanke, H. J., and Raap, A. K. (1991) *In situ* hybridization with fluoresceinated DNA. *Nucleic Acids Res.* **19**, 3237–3241.

3. Ried, T., Baldini, A., Rand, T. C., and Ward, D. C. (1992) Simultaneous visualization of seven different DNA probes by *in situ* hybridization using combinatorial fluorescence and digital imaging microscopy. *Proc. Natl. Acad. Sci. USA* **89**, 1388–1392.

4. Bauman, J. G. J., Wiegant, J., Brost, P., and Van Dujin, P. (1980) A new method for fluorescence microscopical localization of specific DNA sequences by *in situ* hybridization of fluorochrome-labeled RNA. *Exp. Cell Res.* **138**, 485–490.

5. Heslop-Harrison, J. S., Schwarzacher, T., Anamthawat-Jonsson, K., Leitch, A. R., Shi, M., and Leitch. I. J. (1991) *In situ* hybridization with automated chromosome denaturation. *Technique* **3**, 109–116.

6. Schweizer, D. (1980) Counterstain-enhanced chromosomes banding. *Hum. Genet.* **57**, 1–14.

7. Schwarzacher, T. and Heslop-Harrison, J. S. (1991) *In situ* hybridization to plant telomeres using synthetic oligomers. *Genome* **34**, 317–327.

CHAPTER 27

Detection of Digoxigenin-Labeled DNA Probes Hybridized to Plant Chromosomes *In Situ*

Ilia J. Leitch and J. S. (Pat) Heslop-Harrison

1. Introduction

Digoxigenin is widely used as a nonradioactive label for *in situ* hybridization to locate DNA sequences along chromosomes in plants (e.g., *1–3*) and animals. In many cases, its use is similar to biotin (Chapter 25, or e.g., *4*), but digoxigenin labels may give lower unspecific background signal and therefore enable more efficient detection of short or low copy sequences. Digoxigenin is usually incorporated into the DNA enzymatically by nick translation, random priming, polymerase chain reaction (PCR), or end labeling using the modified nucleotide digoxigenin-11-dUTP (*see* Chapters 10 and 11).

For *in situ* hybridization on plant chromosomes, high quality spreads are needed that are free of cytoplasm and cell wall debris (*see* Chapters 23 and 24). The methods for chromosome pretreatments, probe denaturation, and hybridization and posthybridization washes are described in Chapter 26. This chapter describes the probe hybridization mix for digoxigenin-labeled probes (Section 3.1.) and the methods used to detect the sites of probe hybridization (Sections 3.2. to 3.5.).

A variety of different detection systems are available for use with digoxigenin (enzyme-mediated, fluorescence, and colloidal gold) making the digoxigenin system extremely adaptable, versatile, and

From: *Methods in Molecular Biology, Vol. 28:*
Protocols for Nucleic Acid Analysis by Nonradioactive Probes
Edited by: P. G. Isaac Copyright ©1994 Humana Press Inc., Totowa, NJ

sensitive. Detection of digoxigenin-labeled hybrids is mediated via an antidigoxigenin antibody (raised in sheep) that carries the signal generating system; the two most commonly used detection systems for light microscopy are enzyme-mediated detection (Fig. 1A) and fluorescence detection (Fig. 1B).

Enzyme-mediated reporter systems work by catalyzing the precipitation of a visible product at the site of probe hybridization. Antidigoxigenin antibodies are available conjugated to either alkaline phosphatase or horseradish peroxidase (HRP). In this chapter, the enzyme HRP is used to catalyze the oxidation of diaminobenzidine (DAB) to produce a brown precipitate at the site of *in situ* hybridization (Section 3.3.).

Fluorochromes are visualized by excitation with light of one (excitation) wavelength and imaging emitted fluorescence at another (emission) wavelength using appropriate filters. The properties and methods of visualization of the fluorochromes fluorescein and rhodamine, which are used to detect digoxigenin, are the same as for direct fluorochrome-labeled probes and are outlined in Chapter 26.

For the detection of low and single copy sequences the *in situ* hybridization signal can be amplified using antisheep antibodies that also carry the signal generating system (Section 3.2., steps 5–7). The oxidized DAB precipitate generated by the enzyme HRP can be further amplified with silver (Section 3.3., step 4) to increase the sensitivity of detection *(5)*. Sensitivity may also be increased by using the peroxidase antiperoxidase (PAP) complex, which is a preformed soluble enzyme antienzyme immune complex. The greater degree of sensitivity using this complex is mainly attributable to more enzyme molecules being localized per antigenic site.

The particular detection system that is chosen depends on a number of factors. The advantages and disadvantages of each system are outlined below.

Enzyme mediated reporter system:

- Advantages: The *in situ* hybridization signal, which is a precipitate, is stable and can be viewed on a standard light microscope. Since the precipitate can be electron dense (e.g., using HRP and DAB) there is the possibility of transferring the chromosomes into the electron microscope after *in situ* hybridization in order to increase the sensitivity and resolution *(6)*. The sensitivity of detection can also be increased by using silver enhancement (Section 3.3.).

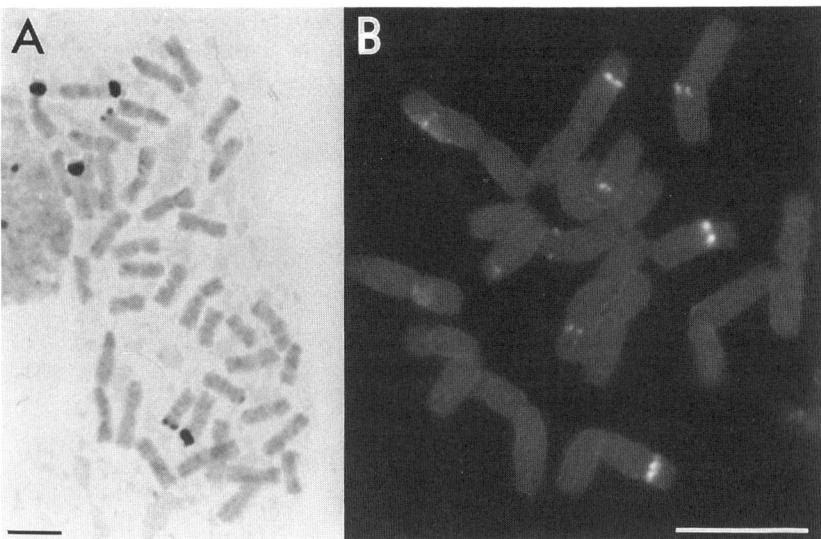

Fig. 1. (**A**) Enzymatic detection of ribosomal DNA on a chromosome spread of wheat (*Triticum aestivum* cv. Chinese Spring). The probe was labeled with digoxigenin and the sites of hybridization detected using horseradish peroxidase conjugated to antidigoxigenin. The sites were visualized by the oxidation of DAB catalyzed by horseradish peroxidase. Four major and two minor sites of probe hybridization are visible. Bar = 10 μm. (**B**) Fluorescent detection of 5S ribosomal DNA sequences on metaphase chromosomes of barley (*Hordeum vulgare*, telotrisomic for the long arm of chromosome 2, 2n = 15). The probe was labeled with digoxigenin and the sites of probe hybridization were detected with fluorescein conjugated to antidigoxigenin. Sites of probe hybridization are visible as bright fluorescent dots on the telosome and eight other chromosomes. Bar = 10 μm.

- Disadvantages: The resolution of the signal is not as high as the fluorescent signal owing to spreading of the precipitate, and the detection of more than one probe simultaneously is more difficult than with fluorochrome-mediated reporter systems. The use of different substrates precipitating different colors has been reported but the distinction of the colors is more difficult than with different fluorochromes.

Fluorochrome-mediated reporter systems:

- Advantages: The signal resolution is high and the simultaneous detection of more than one probe is possible *(1,7)*.
- Disadvantages: An expensive epi-fluorescent microscope is required to visualize the *in situ* hybridization signal and the signal is unstable,

fading on viewing. In addition the signal is unsuitable for visualization by electron microscopy.

In all experiments controls should be run in parallel with experimental slides to determine the specificity of probe hybridization and detection (particularly conducting the experiment without probe and using a known, tested probe).

2. Materials

Unless indicated, all solutions should be made up freshly in deionized or distilled water. All procedures and washes are at room temperature unless otherwise stated. Good laboratory practice is suitable for handling all solutions except DAB, which is carcinogenic (*see* Section 2.3.) and formamide.

2.1. Probe Hybridization Mix for Digoxigenin-Labeled Probes

1. Formamide: Good but not the highest grade (e.g., Sigma [St. Louis, MO] catalog no. F 7508). Store in aliquots at –20°C. Caution this is a teratogen.
2. 50% (w/v) Dextran sulfate in water: Filter (0.22 μm) sterilize and store in aliquots at –20°C.
3. 20X SSC stock: 3M NaCl, 0.3M Na citrate, adjusted to pH 7.0 with HCl, and store at room temperature. Filter (0.22 μm) sterilize.
4. Probe: Digoxigenin-labeled probe is prepared as described in Chapters 10 or 11. The labeled probe is usually dissolved in 1X TE at 20 ng/μL. Incorporation of the digoxigenin label into the DNA should be tested by using a standard dot blot procedure (*see* Note 1).
5. 10% Sodium dodecyl sulfate (SDS) in distilled water: Filter sterilize and store at room temperature.
6. Salmon sperm DNA (5 μg/μL in water): Autoclave to fragments of 100–300 bp (typically 5 min at 100 kgm^{-2}) and store in aliquots at –20°C.
7. Hybridization mix (40 μL/slide; *see* Note 2): For one slide mix (final concentration in brackets), 20 μL formamide (50%), 8 μL 50% dextran sulfate (10%), 4 μL 20X SSC (2X), 4 μL 20 ng/μL digoxigenin-labeled probe (2 ng/μL; *see* Note 3), 1 μL 5 μg/μL salmon sperm DNA (125 ng/μL; *see* Note 4), 2 μL 10% SDS (0.5%; *see* Note 5), 1 μL water.
8. Plastic cover slips: pieces of appropriate size cut from polyethylene autoclavable waste disposal bags.

2.2. Detection of Digoxigenin-Labeled Probes with Antibodies

1. 20X SSC: *See* Section 2.1.
2. Detection buffer: Dilute 20X SSC 1:5 with water and add Tween 20 to 0.2%.

3. Bovine serum albumin (BSA) block: 5% (w/v) BSA (Sigma; Fraction V, globulin-free) in detection buffer.
4. Antidigoxigenin conjugate (raised in sheep; supplied by Boehringer Mannheim, Mannheim, Germany): For enzyme detection use HRP antidigoxigenin (at 7.5 U/mL). For fluorescent detection use fluorescein antidigoxigenin (at 5 µg/mL) or rhodamine antidigoxigenin (at 10 µg/mL). The conjugate is diluted to the recommended concentration in BSA block.
5. Normal rabbit serum block: 5% (v/v) normal rabbit serum in detection buffer.
6. Antisheep conjugate (raised in rabbit; supplied by DAKO [Copenhagen, Denmark]): For enzyme detection use HRP antisheep (at 13 µg/mL). For fluorescent detection use either FITC antisheep (25 µg/mL) or rhodamine antisheep (25 µg/mL) conjugate. The conjugate is diluted to the recommended concentration in normal rabbit serum block.

2.3. Detection of the HRP Reporter (DAB Reaction)

1. DAB detection reagent 1: Prepare 10 mg/mL diaminobenzidine (DAB) in water and store at –20°C in 0.5-mL aliquots. DAB is a carcinogen, so avoid weighing out the powder by ordering small quantities and using the whole vial to make the stock solution. Immediately prior to use dilute one aliquot to 10 mL with 50 mM Tris-HCl, pH 7.4 (final DAB concentration is 0.5 mg/mL). Wear gloves when handling the DAB solution.
2. DAB detection reagent 2: Take 2 mL of DAB detection reagent 1 and add 1 µL 30% hydrogen peroxide solution (Sigma; stored at 4°C). The bottle of hydrogen peroxide solution should be discarded 6 mo after opening.
3. Silver amplification solution A: 0.2% (w/v) ammonium nitrate, 0.2% (w/v) silver nitrate, 1% (w/v) tungstosilicic acid, 0.5% (v/v) formaldehyde (diluted from stock 38% (v/v) formaldehyde in water) in water. Use double-distilled water.
4. Silver amplification solution B: 5% (w/v) Na_2CO_3 in water.
5. 1% (v/v) Acetic acid in water.

2.4. Giemsa Counterstaining of DAB-Treated Chromosomes

1. Sorenson's buffer (pH 6.8): 0.03M KH_2PO_4, 0.03M Na_2HPO_4.
2. Giemsa stain: 4% (v/v) Giemsa solution in Sorenson's buffer. This must be freshly prepared immediately prior to use. If a film appears on the surface remove it with filter paper.
3. Mountant: For temporary mounting use xylene; for permanent mounting use Euparol.

3. Methods

For steps in the schedule using volumes of 200 µL or less the solution is pipeted onto the slide and covered with a plastic cover slip. All washing steps are carried out in Coplin jars using a sufficient volume of liquid to fully immerse the whole slide. Steps are carried out at room temperature unless otherwise indicated.

3.1. In Situ *Hybridization*

Follow the method of *in situ* hybridization as described in Chapter 26 (Sections 2.1. and 3.1. "Chromosome Spread Pretreatment" and Sections 2.2. and 3.2. "Denaturation, Hybridization, and Posthybridization Washes." The hybridization mix containing the digoxigenin-labeled probe (described in Section 2.1. of this chapter) should be substituted at step 2 and 3, Section 3.2.

3.2. Detection
of Digoxigenin-Labeled Probes with Antibodies

Slides should not dry out during washing and detection, and must be handled carefully to avoid scratches. For highly repeated probes only the detection steps (steps 1–4) are necessary; for the detection of low copy sequences the *in situ* hybridization signal can be amplified by performing steps 5–7.

1. Place the slides in detection buffer for 5 min at room temperature.
2. Add 200 µL of BSA block to each slide, cover with a plastic cover slip, and incubate for 5 min at room temperature (*see* Note 6).
3. Remove the cover slip, drain the slide (*see* Note 7), and add 30 µL of the appropriate antidigoxigenin conjugate. Cover and incubate slides in a humid chamber for 1 h at 37°C.
4. Wash slides in detection buffer three times for 5 min each at 37°C. Perform steps 5–7 only when detecting low copy number probes.
5. Add 200 µL of normal rabbit serum block to each slide, cover with a plastic cover slip, and incubate for 5 min at room temperature (*see* Note 6).
6. Remove the cover slip, drain the slide (*see* Note 7), and add 30 µL of the appropriate antisheep conjugate. Cover with a plastic cover slip and incubate slides in a humid chamber for 1 h at 37°C.
7. Wash the slides in detection buffer three times for 8 min at 37°C.

3.3. *Visualization of HRP Reporter (DAB Reaction)*

1. Add 200 µL of DAB detection reagent 1 to each slide, cover with a plastic cover slip, and incubate in the dark at 4°C for 20 min (*see* Note 8).

2. Drain the slides and add 200 μL of DAB detection reagent 2; incubate for a further 20 min in the dark at 4°C (*see* Note 9).
3. Stop the reaction by washing in water three times for 2 min. The signal for low copy sequences can be amplified using silver as follows (*see* Note 10).
4. Immediately prior to use mix an equal volume of silver amplification solution A with solution B and add 500 μL to the slide. At this point the slide may be covered with a glass cover slip and the silver deposition monitored under the microscope (although light exposure hastens silver deposition). Stop the reaction by washing in water followed by 1% acetic acid for 2 min.

3.4. Giemsa Counterstaining of DAB-Treated Chromosomes

1. Examine chromosome preparations to determine whether to counterstain since counterstaining can obscure weak *in situ* hybridization signal.
2. Incubate the slides in Giemsa stain for 5–10 min.
3. Wash slides carefully in a stream of distilled water.
4. Air-dry and mount either temporarily in xylene or permanently in Euparol under a glass cover slip (*see* Note 11). Alternatively the slides can be examined under reflection contrast microscopy (*see* Note 10).
5. Take photographs of areas of interest (*see* Note 12).

4. Notes

1. The probe can be tested by binding a small amount of the labeled probe (0.5–2 μL) to a nitrocellulose or nylon membrane. The incorporation of label is then detected using an enzymic detection system as described in nonradioactive Southern hybridization detection kits (e.g., Boehringer Mannheim or *see* Chapter 18, Section 3.2.).
2. The hybridization mix is usually prepared immediately prior to use but can be stored at –20°C for up to 6 mo.
3. Typically we use between 1 and 5 ng/μL of probe/slide for cloned probes. Increasing the probe concentration tends to result in higher levels of background rather than stronger *in situ* hybridization signal.
4. Blocking DNA is used at concentrations of 2–100X probe concentration. For highly repeated probes we typically use a concentration of 50X the probe concentration.
5. Typically we use 0.2–4 μL of SDS to give a final concentration of 0.05–1%.
6. BSA and normal rabbit serum reduce the level of nonspecific signal by blocking sites, e.g., in the cytoplasm, which might otherwise bind detection reagents.

7. It is important to drain the slides well after this step otherwise the anti-body conjugate is overdiluted. However, the slide must not be allowed to dry out.

8. This step allows the reagents of the enzyme reaction to penetrate evenly into the tissue before the hydrogen peroxide (catalyst) is added.

9. During this stage a precipitate is formed at the site of probe hybridization. Care must be taken not to move the slides during precipitation since this could cause the precipitate to be dislodged and become smeared.

10. The DAB *in situ* hybridization signal can also be visualized by reflection contrast microscopy because the DAB precipitate has very high reflectance properties. This method of visualization can be sensitive, enabling the detection of low levels of *in situ* hybridization signal *(7)*, although cell wall debris gives strong reflectance in plant chromosome preparations.

11. It is suggested that the slide is first mounted in xylene to examine whether the signal needs to be amplified and/or if the counterstain is sufficient. Once this has been determined to be satisfactory, the xylene can be evaporated by placing the slide in an incubator for a few minutes and the cover slip removed. The slide can then be permanently mounted in Euparol.

12. We recommend using color print films as they will record the small differences in color between signal, dirt, and DNA counterstains, while having substantial exposure latitude. We have achieved good results with medium speed films Fujicolor 100 (high contrast), Ektar 125 and Kodak Gold 100. For black and white photography Kodak T-Max 100 or Agfa Ortho professional are suitable.

Acknowledgments

We thank A. R. Leitch, T. Schwarzacher, and K. Anamthawat-Jonsson for sharing their experience with *in situ* hybridization using digoxigenin. IJL thanks AFRC PMB grant 570/111 for support.

References.

1. Leitch, I. J., Leitch, A. R., and Heslop-Harrison, J. S. (1991) Physical mapping of plant DNA sequences by simultaneous *in situ* hybridization of two differently labeled fluorescent probes. *Genome* **34**, 329–333.

2. Maluszynska, J. and Heslop-Harrison, J. S. (1991) Localization of tandemly repeated DNA sequences in *Arabidopsis thaliana. Plant J.* **1**, 159–166.

3. Schwarzacher, T., Anamthawat-Jonsson, K., Harrison, G. E., Islam, A. K. M. R., Jia, J. Z., King, I. P., Leitch, A. R., Miller, T. E., Reader, S. M., Rogers, W. J., Shi, M., and Heslop-Harrison, J. S. (1992) Genomic *in situ* hybridization to identify alien chromosomes and chromosome segments in wheat. *Theor. Appl. Genet.* **84**, 778–786.

4. Schwarzacher, T., Leitch, A. R., Bennett, M. D., and Heslop-Harrison, J. S. (1989) *In situ* localization of parental genomes in a wide hybrid. *Ann. Botany* **64,** 315–324.

5. Manuelidis, L. and Ward, D. C. (1984) Chromosomal and nuclear distribution of the HindIII 1.9 kb human DNA repeat segment. *Chromosoma* **91,** 23–38.

6. Schwarzacher, T. and Heslop-Harrison, J. S. (1991) *In situ* hybridization to plant telomeres using synthetic oligomers. *Genome* **34,** 317–323.

7. Ried, T., Baldini, A., Rand, T. C., and Ward, D. C. (1992) Simultaneous visualization of seven different DNA probes by *in situ* hybridization using combinatorial fluorescence and digital imaging microscopy. *Proc. Nat. Acad. Sci. USA* **89,** 1388–1392.

CHAPTER 28

Preparation of Tissue Sections
and Slides for mRNA Hybridization

Giorgio Terenghi and Julia M. Polak

1. Introduction

The first step for a successful *in situ* hybridization is the fixation of
the tissue. This will ensure target nucleic acid retention and preservation
of the tissue morphology. Either crosslinking or precipitative fixatives
can be used, and a preference for either of the two types of fixative
has often been based on the different types of system under investigation *(1–7)*. For hybridization of regulatory peptide mRNA, 4%
paraformaldehyde appears to be the most effective, both on tissue blocks
and on tissue culture preparations.

When manipulating tissue to be used for mRNA hybridization, it is
essential to work in RNase-free conditions. RNase is an ubiquitous and
heat-resistant enzyme that degrades any single-stranded RNA molecule
very rapidly. Fingertips are particularly rich in ribonuclease, hence
clean, disposable gloves should be worn at all times. All equipment and
solutions should also be RNase-free. Fixative solutions, as they exert
an inhibitory action on this enzyme, are naturally RNase-free.

Fixed tissue is generally processed for cryostat sectioning, but *in
situ* hybridization can be equally successful on paraffin embedded
tissue *(8,9)*. In any case, it is important always to keep the delay
between tissue collection and fixation to a minimum, to avoid nucleic
acid degradation, which obviously increases with time delay *(4,10)*. This

From: *Methods in Molecular Biology, Vol. 28:*
Protocols for Nucleic Acid Analysis by Nonradioactive Probes
Edited by: P. G. Isaac   Copyright ©1994 Humana Press Inc., Totowa, NJ

is not a problem when using experimental animal tissue or cell culture, but it becomes an extremely important point when using surgical or postmortem material. As the degradation curve varies for different mRNAs, there is no fixed rule on an acceptable time limit, but a delay of 15–30 min is considered acceptable in most cases.

The fixed material can be stored in washing buffer only for a limited time (up to 1 mo), but frozen or paraffin tissue blocks can be safely stored for many months or years. Cryostat blocks should be stored at –40°C or below, and room temperature is considered adequate for paraffin blocks. Cryostat sections can be stored dry at –70°C for up to 1 yr, and wax sections keep at room temperature (dewaxed sections can be kept in 70% alcohol at 4°C), given that RNase-free conditions are observed.

Tissue sections should be collected on poly-L-lysine (PLL) coated slides to prevent loss of material during the many steps of the hybridization procedure. The best tissue adhesion is obtained if the sections are left to dry for at least 4 h (or overnight) at 37°C before use or storage.

2. Materials

Plastic disposable equipment and solutions should be autoclaved before use. RNase inhibitors (i.e., human placental ribonuclease inhibitor, DEPC, heparin, and so on) can be added to solutions containing enzymes, which are heat sensitive.

1. Phosphate buffered saline (PBS): Dissolve in 9 L of distilled water the following: 87.9 g NaCl, 2.72 g KH_2PO_4, 11.35 g anhydrous Na_2HPO_4. Adjust the pH to 7.1–7.2 with HCl before adjusting the total volume to 10 L. This solution can be stored at room temperature ready for use.
2. 4% Paraformaldehyde solution: Dissolve 4 g of paraformaldehyde in 80 mL of 0.01M PBS with heat, keeping the temperature below 60°C. Stir the slurry until the powder is completely dissolved, and if necessary add few drops of 10N NaOH to clear the solution. Adjust the volume to 100 mL and leave the solution to cool before using. The fixative should be freshly prepared before use.
3. Poly-L-lysine (PLL) solution: Dissolve PLL in sterile distilled water (1 mg/mL). The optimum mol wt of PLL is >300,000. The solution may be stored frozen at –20°C in small aliquots. Thawing and refreezing is not harmful, but thawed solutions must be mixed well before use.

4. RNase-free glass slides: Place some slides in metal racks and soak them in 0.2% Triton X-100 overnight. Rinse the slides in running tap water for about 1 h, then rinse in distilled water. Dry the slides and racks at 37–50°C. When dry, wrap them in aluminum foil and bake in an oven at 250°C for at least 4 h, preferably overnight.
5. Liquid nitrogen.
6. Arcton 12 (dichlorodifluoromethane) (ICI, UK), or isopentane.
7. Freezing glue (e.g., Tissue-Tek OCT, Miles Inc., USA).
8. Washing buffer: 15% (w/v) sucrose, 0.01% (w/v) sodium azide dissolved in PBS.
9. PLL-coated slides: Always wear gloves during this procedure. Place the RNase-free slide on clean aluminum foil or a similar surface that has been wiped with absolute alcohol to avoid RNase contamination. With a pencil, mark the slide on the side to be coated, as PLL is transparent and impossible to detect once applied. Apply a small drop (approx 5 μL) of PLL solution to the slides. Spread a thin film of PLL over the whole surface. Either use the same method as for making blood films, or appose the surface of two slides to spread the drops of liquid and then slide them apart. The PLL film dries quickly (1–2 min) and the tissue sections can be picked up onto the coated surface immediately afterwards (*see* Note 1).

3. Methods

3.1. Fixation (see Note 2)

1. Cut tissue into small pieces (approx $1 \times 1 \times 0.5$ cm) using a sterile sharp blade (*see* Note 3).
2. Fix the tissue in freshly made 4% paraformaldehyde solution for 6 h at room temperature (*see* Note 4).
3. After fixation, rinse the tissue blocks in four to five changes of washing buffer (2 h each change or overnight).
4. Store the fixed material in washing buffer at 4°C in labeled containers, ready for cryostat or wax blocking (*see* Note 5).

3.2. Preparing the Cryostat Block and Sectioning Tissue

1. Precool a Pyrex or metal beaker by immersion in liquid nitrogen and fill it with Arcton or isopentane.
2. Immerse the beaker again in liquid nitrogen and freeze the Arcton (isopentane), then remove the beaker from the flask and allow the Arcton (isopentane) to melt until there is enough liquid to cover the block, but

still some solid in the bottom to maintain the temperature as low as possible.

3. Mount the tissue on a cork disk and surround it with special freezing glue (e.g., OCT) (*see* Notes 6 and 7).
4. Hold the cork disk with forceps and lower it quickly in the melting Arcton (isopentane) to snap-freeze the tissue.
5. Transfer the frozen block in precooled plastic bag for storage in liquid nitrogen or at –40°C.
6. The block should be allowed to warm up to cryostat temperature (–20°C) before cutting is attempted.
7. Mount the block on the cryostat head and cut thin sections (10–30 μm), picking them up onto PLL-coated slides (*see* Notes 1, 5, and 7).
8. Dry the sections for at least 4 h (or overnight) at 37°C to obtain maximum tissue adhesion.

4. Notes

1. PLL-coated slides are best used immediately, but batches may be prepared and stored in the baked rack, wrapped in the aluminum foil to protect from dust and RNase contamination. PLL-coated slides can be stored at room temperature for up to 1 mo.
2. When mRNA is the target, special care should be taken, and all the specimen handling procedures should be carried out using clean, disposable gloves and sterile instruments, in order to avoid RNase contamination.
3. Animal tissue can be fixed *in situ* by perfusion with 4% paraformaldehyde, followed by immersion fixation of the dissected tissue (1–4 h, depending on the tissue size and fixation obtained with perfusion). This method is strongly recommended if brain or spinal cord tissues are used, as these tissues do not fix well by immersion only, owing to the poor penetration of the fixative in the tissue matrix.
4. Fixative other than paraformaldehyde may be more appropriate when investigating specific target molecules. Some indication of other possible fixatives can be gained from the listed literature. However, it is good practice to test more than one fixative in order to establish which give best hybridization results, as indicated by the highest signal:background noise ratio with optimal preservation of morphology.
5. During storage of the cryostat blocks, do not leave any tissue surface uncovered, as it will dry out and it will become impossible to cut. After cutting, spread a thin layer of OCT glue on the cut surface of the block, and leave at –20°C until frozen. Store the block in a sealed plastic bag or other appropriate container.

6. Tissue blocks should be orientated so that the face you wish to cut is uppermost. Cryostat blocks cut best in the vertical plane, unlike paraffin blocks.
7. Very small pieces of tissue should be mounted on another piece of inert or inappropriate tissue (e.g., liver), suitably trimmed, so that adequate clearance is obtained on cutting.

References

1. Haase, A. T., Brahic, M., and Stowring, L. (1984) Detection of viral nucleic acids by *in situ* hybridization, in *Methods in Virology*, vol. VII (Maramorosch, K. and Koprowski, H., eds.), Academic, New York, pp. 189–226.
2. McAllister, H. A. and Rock, D. L. (1985) Comparative usefulness of tissue fixatives for *in situ* viral nucleic acid hybridization. *J. Histochem. Cytochem.* **33,** 1026–1032.
3. Moench, T. R., Gendelman, H. E., Clements, J. E., Narayan, O., and Griffin, D. E. (1985) Efficiency of *in situ* hybridization as a function of probe size and fixation technique. *J. Virol. Method.* **11,** 119–130.
4. Hofler, H., Childers, H., Montminy, M. R., Lechan, R. M., Goodman, R. H., and Wolfe, H. J. (1986) *In situ* hybridization methods for the detection of somatostatin mRNA in tissue sections using antisense RNA probes. *Histochem. J.* **18,** 597–604.
5. Singer, R. H., Lawrence, J. B., and Villnave, C. (1986) Optimization of *in situ* hybridization using isotopic and non-isotopic detection methods. *Biotechniques* **4,** 230–250.
6. Guitteny, A. F., Fouque, B., Mongin, C., Teoule, R., and Boch, B. (1988) Histological detection of mRNA with biotinylated synthetic oligonucleotide probes. *J. Histochem. Cytochem.* **36,** 563–571.
7. Terenghi, G. and Fallon, R. A. (1990) Techniques and applications of *in situ* hybridization, in *Current Topics in Pathology: Pathology of the Nucleus* (Underwood, J. C. E., ed.), Springer Verlag, Berlin, pp. 290–337.
8. Farquharson, M., Harvie, R., and McNicol, A. M. (1990) Detection of mRNA using a digoxigenin end labelled oligodeoxynucleotide probe. *J. Clin. Pathol.* **43,** 424–428.
9. Unger, E. R., Hammer, M. I., and Chenggis, M. L. (1991) Comparison of 35S and biotin as labels for *in situ* hybridization: use of an HPV model system. *J. Histochem. Cytochem.* **39,** 145–150.
10. Asanuma, M., Ogawa, N., Mizukawa, K., Haba, K., and Mori, A. (1990) A comparison of formaldehyde-preperfused frozen and freshly frozen tissue preparation for the *in situ* hybridization for α-tubulin mRNA in the rat brain. *Res. Comm. Chem. Pathol. Pharmacol.* **70,** 183–192.

CHAPTER 29

Detecting mRNA in Tissue Sections with Digoxigenin-Labeled Probes

Giorgio Terenghi and Julia M. Polak

1. Introduction

Nonradioactively labeled probes offer several advantages compared to radioactive ones, as they show long stability, high morphological resolution, and rapid developing time. There are different types of nonradioactive labeling methods available, although recently digoxigenin-labeled probes *(1)* have become the most widely used for investigation on animal tissue, as they offer the advantage of low background noise and increased sensitivity *(2,3)*. Also, digoxigenin can be used to label either RNA, DNA, or oligonucleotide probes. There have been different opinions on the sensitivity of detection of digoxigenin probes, but recent publications have shown that the sensitivity of radiolabeled and nonradioactive probes is comparable *(3)*.

The detection of digoxigenin-labeled probes is carried out with immunohistochemical methods, using antidigoxigenin antibodies that are conjugated to either fluorescent or enzymatic reporter molecules *(1)*. However, it has to be remembered that different immunohistochemical detection systems might determine the resolution and detection sensitivity of *in situ* hybridization *(4)*. The variety of detection methods also offers the possibility to carry out double *in situ* hybridization, e.g., using digoxigenin-labeled probes and directly labeled probes, which are then visualized using different immunohistochemical methods *(5)*.

From: *Methods in Molecular Biology, Vol. 28:*
Protocols for Nucleic Acid Analysis by Nonradioactive Probes
Edited by: P. G. Isaac Copyright ©1994 Humana Press Inc., Totowa, NJ

Alternatively, digoxigenin and biotin can be used in combination for the identification of two different nucleic acid sequences on the same sections *(6,7)*.

This chapter describes the use of digoxigenin-labeled probes to detect mRNA transcripts in tissue sections. The sites of hybridization of the probe are visualized using an antidigoxigenin antibody:alkaline phosphatase conjugate, and a colorimetric reaction.

2. Materials

1. $1M$ Tris-HCl, pH 8: Dissolve 121.1 g Tris-base in 800 mL double-distilled water. Adjust to pH 8 with HCl, then adjust volume to 1 L with double-distilled water before autoclaving.
2. $0.5M$ EDTA: Add 186.1 g of $Na_2EDTA \cdot 2H_2O$ to approx 600 mL double-distilled water. Stir continuously keeping the solution at 60°C, adding NaOH pellets (approx 20 g) until near pH 8. Only then will the EDTA start to dissolve. When completely dissolved, leave the solution to cool down to room temperature, then adjust to pH 8 with $10N$ NaOH solution. Adjust the volume to 1 L with double-distilled water and autoclave.
3. $1M$ Glycine: Dissolve 75 g glycine in 800 mL double-distilled water. When dissolved, adjust volume to 1 L and autoclave.
4. $1M$ Triethanolamine: Mix 44.5 mL triethanolamine in 200 mL double-distilled water, adjust to pH 8 with HCl, then bring to 300 mL volume with double-distilled water before autoclaving.
5. 10X SSC (standard saline citrate): Dissolve 87.65 g NaCl and 44.1 g sodium citrate in 800 mL double-distilled water, adjust to pH 7 with $10N$ NaOH, then bring to 1 L volume. Autoclave a small aliquot to be used for the hybridization buffer. When the solution is used for posthybridization washes, it does not need to be autoclaved (*see* Note 1).
6. Deionized formamide: Mix 50 mL of formamide and 5 g of mixed-bed ion-exchange resin (e.g., Bio-Rad AG 501-X8, 20–50 mesh). Stir for 30 min at room temperature, then filter twice through Whatman No.1 filter paper. Store in small aliquots at –20°C.
7. 100X Denhardt's solution: Dissolve 1 g Ficoll, 1 g polyvinylpyrrolidone, and 1 g bovine serum albumin (BSA) (Fraction V) in 50 mL of sterile distilled water and autoclave.
8. Herring sperm DNA: Dissolve the DNA (Type XIV sodium salt) in sterile distilled water at a concentration of 10 mg/mL. If necessary, stir the solution on a magnetic stirrer for 2–4 h at 37–40°C to help the DNA to dissolve. Shear the DNA by passing it several times through a sterile 18-g hypodermic needle. Alternatively sonicate on medium-high power for 5 min. Boil the DNA solution for 10 min and store at –20°C in small

aliquots. Just before use, heat the DNA for 5 min in a boiling water bath. Chill it quickly in ice-water.

9. 10% Sodium dodecyl sulfate (SDS): Dissolve 100 g SDS in 900 mL of double-distilled water. Heat to 68°C to assist solubilization. Adjust to pH 7.2 by adding a few drops of HCl. Adjust volume to 1 L. This solution does not need autoclaving, as it is an RNase inhibitor.

10. Hybridization buffer: 50% deionized formamide, 5X SSC, 10% dextran sulfate, 5X Denhardt's solution, 2% SDS, and 100 µg/mL denatured sheared herring sperm DNA. Make this solution fresh before use, and store at 50°C.

11. Phosphate buffered saline (PBS): dissolve in 9 L of distilled water the following: 87.9 g NaCl, 2.72 g KH_2PO_4, 11.35 g Na_2HPO_4 anhydrous. Adjust the pH to 7.1–7.2 with HCl before adjusting the total volume to 10 L. Autoclave before use.

12. 0.2% Triton-X100 (v/v) in autoclaved PBS.

13. Proteinase K (stock solution): Dissolve the proteinase K at 0.5 mg/mL concentration in sterile distilled water. Divide into small aliquots and store at –20°C.

14. Permeabilizing solution: Prepare $0.1M$ Tris-HCl, pH 8.0, 50 mM EDTA from stock solutions ($1M$ and $0.5M$ respectively), diluting 1/10 with sterile distilled water. Just before use (Section 3.1., step 2), to 100 mL of this solution, prewarmed at 37°C, add 200 µL proteinase K stock solution (final concentration 1 µg/mL).

15. $0.1M$ Glycine in PBS: Dilute $1M$ stock solution 1/10 with autoclaved PBS.

16. 4% Paraformaldehyde in PBS: Dissolve 4 g of paraformaldehyde in 80 mL of $0.01M$ PBS with heat, keeping the temperature below 60°C. Stir the slurry until the powder is completely dissolved, and if necessary add few drops of 10N NaOH to clear the solution. Adjust the volume to 100 mL and leave the solution to cool before using. The fixative should be freshly prepared before use.

17. Acetic anhydride.

18. $0.1M$ Triethanolamine: freshly prepared 1/10 dilution of $1M$ stock solution in sterile distilled water.

19. Sterile double-distilled water.

20. Digoxigenin-labeled cRNA probe (*see* Chapter 12). The probe should be complementary to the target mRNA. The probe should be ethanol precipitated and dissolved in hybridization buffer immediately before use (Section 3.1., step 8).

21. 5X SSC (*see* Note 1).

22. 2X SSC, 0.1% SDS (*see* Note 1).

23. 0.1X SSC, 0.1% SDS (*see* Note 1).

24. 2X SSC (*see* Note 1).

25. RNase stock solution: Dissolve pancreatic RNase (RNase A) at a concentration of 10 mg/mL in 10 m*M* Tris-HCl, pH 7.5, and 15 m*M* NaCl. Dispense into aliquots and store at –20°C.
26. 10 μg/mL RNase in 2X SSC: Add 100 μL of RNase stock solution to 100 mL 2X SSC prewarmed at 37°C. Prepare freshly before use.
27. Sections on slides (*see* Chapter 28).
28. Buffer 1: 0.1*M* Tris-HCl, pH 7.5, 0.1*M* NaCl, 2 m*M* MgCl$_2$, 3% BSA. In 800 mL double-distilled water dissolve 12.1 g Tris-base and 5.85 g NaCl. Adjust to pH 7.5, then add 0.2 g MgCl$_2$. Adjust the volume to 1 L and add BSA to 3%.
29. Buffer 2: 0.1*M* Tris-HCl, pH 9.5, 0.1*M* NaCl, 50 m*M* MgCl$_2$. In 800 mL double-distilled water dissolve 12.1 g Tris-base and 5.85 g NaCl. Adjust to pH 9.5, then add 4.4 g MgCl$_2$. Adjust the volume to 1 L.
30. Antidigoxigenin antibody/alkaline phosphatase conjugate (Boerhinger): Immediately before use (Section 3.2., step 3) dilute to 1/500 with buffer 1.
31. Nitroblue tetrazolium chloride (NBT): Dissolve 35 mg NBT in 277 μL 70% dimethylformamide (DMF). Prepare freshly before use.
32. 5-Bromo-4-chloro-3-indolyl-phosphate (BCIP): Dissolve 17 mg BCIP in 222 μL 100% DMF. Prepare freshly before use.
33. Substrate buffer: In 100 mL buffer 2 dissolve 25 mg levamisole (Sigma, St. Louis, MO), then add 277 μL NBT solution and 222 μL BCIP solution just before use.
34. Stop buffer: 20 m*M* Tris-HCl, pH 7.5, 5 m*M* EDTA. In 800 mL double-distilled water dissolve 2.42 g Tris-base, then adjust to pH 7.5 with HCl. Add 10 mL 0.5*M* EDTA stock solution and adjust volume to 1 L.
35. 5% Pyronin Y: Dissolve 5 g pyronin Y in 100 mL double-distilled water.

3. Methods

All the steps up to hybridization (included) should be carried out in RNase-free conditions. Solutions should be autoclaved or prepared with sterile ingredients using RNase-free equipment. Equipment should be autoclaved, or baked at 250°C for 4 h, as appropriate.

Select the slides, number, and mark them as necessary with pencil (not pen, ink may disappear during the various incubation steps).

3.1. Hybridization

1. Rehydrate the sections by immersion in 0.2% Triton/PBS for 15 min. Wash in PBS twice for 3 min.
2. Carry out the tissue permeabilization by incubating the tissue in permeabilizing solution prewarmed at 37°C, containing 1 μg/mL proteinase K. The normal incubation time is 15–20 min (*see* Note 2).

3. Stop the proteinase K activity by immersion in 0.1M glycine in PBS for 5 min.

4. Immerse the sections in 4% paraformaldehyde for 3 min to postfix the target nucleic acid.

5. Rinse the sections briefly in PBS, twice, to remove the paraformaldehyde.

6. Place the slides in a staining jar containing 0.1M triethanolamine and, while stirring, add acetic anhydride to 0.25% (v/v) and incubate for 10 min (*see* Note 3).

7. Rinse the slides briefly in double-distilled water and dry them at 37–40°C. This takes approx 10 min.

8. Dissolve the probe in hybridization buffer at 50°C to a final concentration of 2.5 ng/μL.

9. Apply 10 μL of diluted probe per section to the dry slides. The volume of diluted probe can be increased for large sections.

10. Using fine forceps, gently place a siliconized cover slip onto the section to spread the probe solution. If there are any air bubbles, remove them by pressing gently on the cover slip with the forceps.

11. Hybridize the section for 16–20 h (*see* Note 4) at suitable temperature in a sealed humid chamber containing 5X SSC. A different hybridization temperature will be needed for various probes according to their T_m (*see* Note 5).

12. Following hybridization, remove the cover slip by immersing the slide in 2X SSC, 0.1% SDS. The cover slips will float off after few minutes soaking.

13. Wash the sections in 2X SSC, 0.1% SDS at room temperature, shaking gently, for four changes of 5 min.

14. Wash the slides in 0.1X SSC, 0.1% SDS at the same temperature used for hybridization, shaking gently, for two changes of 10 min (*see* Note 6).

15. If using cRNA probes, rinse the sections briefly in 2X SSC, twice, then incubate 10 μg/mL RNase A solution in 2X SSC at 37°C for 15 min (*see* Note 7).

16. Rinse briefly in 2X SSC, then PBS before proceeding with the immunohistochemistry detection.

3.2. Immunohistochemistry Detection (see Note 8)

1. Block nonspecific binding by immersing the slides in buffer 1 for 10 min at room temperature.

2. Wipe dry the slides around the tissue but keep the tissue wet.

3. Put on the section a drop of antidigoxigenin antisera conjugated to alkaline phosphatase, diluted 1/500 with buffer 1. Incubate 2 h at room temperature (*see* Notes 8 and 9).

4. Wash in buffer 1 for three changes of 3 min.
5. Equilibrate the sections in buffer 2 for 10 min at room temperature.
6. Immerse the slides in substrate buffer for 10–30 min (or longer if necessary) at room temperature, covering the dish with aluminum foil to keep the reaction in the dark (*see* Note 10).
7. Check the slides under the microscope to assess the development reaction. Stop the reaction or put the slides back in the solution and leave for a longer time as required. The color of the reaction is blue-black.
8. Stop the reaction by immersing in stop buffer for 5 min at room temperature.
9. Counterstain by dipping the slides in 5% pyronin Y solution, for 10–30 s.
10. Rinse well under tap water, approx 5–10 min, until the water is clear.
11. While still wet, mount the slides in aqueous mountant (e.g., Hydromount or similar).

4. Notes

1. There is no need to autoclave the solutions used for posthybridization washes.
2. The incubation time for proteinase K should be titrated for tissue type, as prolonged proteinase digestion could damage the tissue, with a loss of morphology and of target nucleic acid.
3. The treatment with triethanolamine is carried out in order to acetylate the tissue. This prevents electrostatic interaction between the tissue and the probe, as a result of opposite electrostatic charges, thus reducing the background staining.
4. It has been demonstrated that the hybridization reaction reaches an equilibrium after 4–6 h incubation, when a maximum of hybrid has formed. However, incubation is generally carried out overnight for convenience.
5. The hybridization temperature is dependent on the type of probes that have been used. With hybridization buffer containing 50% formamide, it is suggested that the following range of temperature should be tested initially: 42–48°C for cDNA probes; 42–55°C for cRNA probes; 37–40°C for oligonucleotide probes.
6. High background signal can be removed by prolonged washes of the slides at higher stringencies. Increase the number of washes in 0.1X SSC, 0.1% SDS, also progressively increase the temperature. Take care not to allow the sections to dry out in between any of the washes. Use the same container throughout, changing the solutions quickly.
7. It is essential to remove any trace of SDS from the sections before the incubation with RNase solution, as SDS would inhibit the action of the enzyme. It is desirable to use RNase at this stage of the hybridization procedure if you have been using cRNA probes. RNaseA will degrade

the single-stranded cRNA probe that is bound nonspecifically to the section, hence decreasing the background staining. Double-stranded RNA hybrids (cRNA-mRNA) are unaffected by the enzyme.

8. There is a variety of reporter molecules (e.g., alkaline phosphatase, FITC, and so on) conjugated to antibodies against digoxigenin, available from Boehringer, which may be suitable for different applications. There are also detection kits available from the same supplier, which include all reagents needed for the procedure. It is suggested that the reader refers to the catalog from Boehringer for further details.

9. Antibodies conjugated to other reporter molecules may require different dilutions, which are specified by the supplier.

10. The developing time varies considerably according to the type of tissue and the abundance of the target nucleic acid within the cell. In some cases several hours, or overnight, incubation in substrate buffer is necessary to obtain detectable signal. However, it has to be remembered that prolonged incubation also increases the nonspecific background staining.

References

1. Kessler, C. (1991) The digoxigenin anti-digoxigenin (DIG) technology—A survey on the concept and realization of a novel bioanalytical indicator system. *Mol. Cell. Probes* **5,** 161–205.

2. Morris, R. G., Arends, M. J., Bishop, P. E., Sizer, K., Duvall, E., and Bird, C. C. (1990) Sensitivity of digoxigenin and biotin labeled probes for detection of human papillomavirus by *in situ* hybridization. *J. Clin. Pathol.* **43,** 800–805.

3. Furuta, Y., Shinohara, T., Sano, K., Meguro, M., and Nagashima, K. (1990) *In situ* hybridization with digoxigenin-labeled DNA probes for detection of viral genomes. *J. Clin. Pathol.* **43,** 806–809.

4. Giaid, A., Hamid, Q., Adams, C., Springall, D. R., Terenghi, G., and Polak, J. M. (1989) Non-isotopic RNA probes. Comparison between different labels and detection systems. *Histochemistry* **93,** 191–196.

5. Dirks, R. W., van Gijlswijk, R. P. M., Tullis, R. H., Smit, A. B., van Minnen, J., van der Ploeg, M., and Raap, A. K. (1990) Simultaneous detection of different mRNA sequences coding for neuropeptide hormones by double *in situ* hybridization using FITC- and biotin-labeled oligonucleotides. *J. Histochem. Cytochem.* **38,** 467–473.

6. Herrington, C. S., Burns, J., Graham, A. K., Bhatt, B., and McGee, J. O. D. (1989) Interphase cytogenetics using biotin and digoxigenin labeled probes. II: Simultaneous detection of two nucleic acid species in individual nuclei. *J. Clin. Pathol.* **42,** 601–606.

7. Trask, B. J., Massa, H., Kenwrick, S., and Gitschier, J. (1991) Mapping of human chromosome Xq28 by two colour fluorescence *in situ* hybridization of DNA sequences to interphase cell nuclei. *Am. J. Hum. Genet.* **48,** 1–15.

CHAPTER 30

Detection of mRNA in Whole Mounts of Mouse Embryos Using Digoxigenin Riboprobes

Barry Rosen and Rosa Beddington

1. Introduction

This chapter describes a nonradioactive method for the localization of mRNA in whole mouse embryos. It employs riboprobes labeled with digoxigenin, a steroid-like moiety not found in animal tissue. Digoxigenin-containing probe is visualized with a conjugate of antidigoxigenin Fab and alkaline phosphatase and colorimetric staining. The results are visualized in three dimensions, hence subtle patterns can be visualized without laborious sectioning.

The procedure is rapid, yielding results in 3 d. It generally gives good signal-to-noise ratios and resolution at the single cell level. The method has been used on 5.5–10 d postcoitum embryos and also on blastocysts. It has been used to detect a number of RNA sequences with various localizations in the embryo, including actin, Oct-4, Int-2, Krox-20, and the T(*Brachyury*) gene product *(1)*.

This protocol is derived from techniques first developed for *Drosophila* embryos *(1)*. The chief difference is the use of ionic and nonionic detergents as opposed to proteinases to increase hybridization and detection efficiency. We have found that the use of proteases was the major cause of experimental variability, both between different experiments and within different stages and regions of embryos. We

From: *Methods in Molecular Biology, Vol. 28:*
Protocols for Nucleic Acid Analysis by Nonradioactive Probes
Edited by: P. G. Isaac Copyright ©1994 Humana Press Inc., Totowa, NJ

have also endeavored to reduce the numbers of treatments and washes
to an absolute minimum, as these result in the loss and/or destruction
of embryos.

Probing of whole embryo mounts and subsequent detection by an
antibody/alkaline phosphatase conjugate presents two major problems. The probe and antibody conjugate can become trapped in cavities
in the embryo. For this reason controls should include embryos probed
with a heterologous probe, and embryos processed in the absence of
probe. Secondly, endogenous alkaline phosphatase in the tissue can
give rise to spurious positive signals. The procedure described in this
chapter takes steps to minimize the likelihood of detecting endogenous
alkaline phosphatase (by using a combination of low pH and high
temperature during the hybridization). However, controls should also
be included of embryos that have neither probe nor antibody added.

2. Materials

Diethylpyrocarbonate(DEPC) treated solutions are prepared by
adding DEPC to 0.1% vol (in a fume-hood), mixing vigorously, and
incubating either overnight at room temperature or for 2 h at 37°C.
DEPC solutions are then autoclaved for 20 min to inactivate DEPC.

1. Medium containing 10% fetal calf serum.
2. PBS (phosphate buffered saline): (Sigma [St. Louis, MO] tablets) DEPC-
   treated.
3. 4% Paraformaldehyde in PBS. Add the paraformaldehyde powder to
   the PBS and heat the solution a few hours at 80°C to dissolve. Use the
   same day.
4. PBT: PBS containing 0.1% Tween-20. Mix vigorously to dissolve.
5. 25% Methanol: 25% methanol in PBS. Store at room temperature ready
   for use.
6. 50% Methanol : 50% methanol in PBS. Store at room temperature ready
   for use.
7. 75% Methanol : 75% methanol in PBS. Store at room temperature ready
   for use.
8. 100% Methanol. Store at room temperature.
9. 10-mL Cylindrical glass vials (Reacti-Vials) manufactured by Pierce
   Ltd. Vials should be silanized, acid-washed, and DEPC-treated (*see* Note
   1). 20-mL plastic conical vials with screw cap (Sarstedt) can also be
   used. These should be autoclaved before use.

10. RIPA: 150 mM NaCl, 1% NP-40, 0.5% sodium deoxycholate (DOC), 0.1% sodium dodecyl sulfate, 1 mM EDTA, 50 mM Tris-HCl, pH 8.0. This mixture is made from autoclaved stock solutions of everything except DOC and NP-40. Stock solutions of these last two may be made with autoclaved water.

11. 4% Paraformaldehyde, 0.2% glutaraldehyde in PBT *(see above)*. Make this solution from a frozen, E. M. grade 25% glutaraldehyde stock (Sigma Grade 1) and freshly prepared paraformaldehyde.

12. 20X SSC: 3M NaCl, 0.3M trisodium citrate, pH 4.5. Adjust pH with 1.0M citric acid. Treat with DEPC before using.

13. Hybridization buffer: For 10 mL mix 5.0 mL ultrapure (Gibco/BRL, Gaithersburg, MD) or deionized formamide, 2.5 mL DEPC-treated 20X SSC pH 4.5, 5 µL 100 mg/mL Heparin in DEPC water, 10 µL Tween-20, and 2.5 mL DEPC-treated water.

14. 1:1 Hybridization buffer: PBT.

15. Hybridization buffer containing 100 µg/mL tRNA and 100 µg/mL sheared, denatured herring sperm DNA. This solution should be made from stock solutions just before use. The RNA and DNA should both be phenol extracted, and should be stored at –20°C. The DNA can be sheared by sonication or passaging repeatedly through a fine-gage hypodermic needle. Immediately before the solution is made, denature the DNA by heating in a boiling water bath for 5 min, then plunge the DNA into iced water.

16. Digoxigenin-labeled RNA probe *(see* Note 2). Immediately before use denature the probe by heating to 80°C for several minutes, cool it on ice, and then add the hybridization solution to the probe *(see* Section 3.3.)

17. 2X SSC pH 4.5, 50% formamide, 0.1% Tween-20.

18. 1X TBST: This is made by diluting a 10X stock. For 100 mL of 10X TBST stock, mix 8 g NaCl, 0.2 g KCl, 25 mL 1M Tris-HCl, pH 7.5, 10 mL 10% Tween-20. Add DEPC-treated water to 100 mL.

19. 10% Heat-inactivated sheep serum in TBST. The inactivation of the serum is carried out at 70°C for 30 min, before mixing with the TBST. This solution should be freshly prepared before use.

20. Acetone embryo powder: This is made from 11.5–12.5 d embryos by acetone extraction using standard techniques. Embryos are homogenized thoroughly on ice in PBS at the ratio of 1 g of tissue/1.0 mL PBS. Four volumes of acetone (–20°C) are then added. Mix well and store on ice for at least 30 minutes. Centrifuge at 10,000g for 30 min and discard the supernatant. Resuspend the pellet in acetone (–20°C). Store on ice at least 10 min. Spin as before and discard supernatant. Transfer pellet

to a clean piece of filter paper and air-dry at room temperature. Spread and disperse the precipitate as it dries to make a powder. Store powder in freezer at −20°C.

21. Preadsorbed diluted antibody conjugate: This is made up just before use during the washing and blocking steps (Section 3.4.). Heat inactivate approx 5 mg of acetone embryo powder in 1X TBST for 30 min at 70°C. Spin quickly (30 s in microfuge) to settle powder and remove supernatant. Cool the pellet on ice. Add a 1:500 dilution of antidigoxigenin Fab/alkaline phosphatase conjugate (Boehringer Mannheim, Mannheim, Germany) in 1% heat inactivated sheep serum/TBST. Disperse the powder well and incubate with gentle shaking for 1 h at 4°C. Spin for 5 min at 4°C and transfer the supernatant to a fresh tube. Dilute the supernatant fourfold with 1% heat-inactivated sheep serum in TBST to a final antibody dilution of 1:2000.

22. Alkaline phosphatase buffer: 100 mM NaCl, 50 mM MgCl$_2$, 0.1% Tween-20, 100 mM Tris-HCl, pH 9.5. This solution should be made fresh from stock solutions.

23. NBT solution: 75 mg/mL nitroblue tetrazolium in 70% dimethylformamide, 30% water (Gibco/BRL).

24. BCIP(X-Phosphate) solution: 5-bromo-4-chloro-3-indolyl phosphate, toluidinium salt, 50 mg/mL in dimethylformamide (Gibco/BRL).

25. Staining solution: This solution is made up just before use by mixing 4.5 μL NBT solution and 3.5 μL BCIP (X-Phosphate) solution/mL of alkaline phosphatase buffer (above).

26. PBT containing 1 mM EDTA.

3. Methods

3.1. Dissection and Fixation

1. Dissect the embryos (*see* Note 3) at room temperature in PB1 medium containing 10% fetal calf serum. Dissect away as many extraembryonic membranes as possible, as they can trap probe/antibody leading to high background.

2. Place the embryos on ice in PBS.

3. Wash the embryos once or twice with ice-cold PBS.

4. Fix embryos in freshly prepared 4% paraformaldehyde in PBS at 4°C. Fixation time of 2–5 h is sufficient, but overnight fixation works well, too, and can be more convenient.

5. Wash the embryos twice on ice with PBT.

6. Dehydrate the embryos by passaging (on ice) for 5–10 min each through 25, 50, and 75% methanol, and twice in 100% methanol (*see* Note 4).

7. Store the embryos at −20°C (*see* Note 5).

3.2. Permeabilization and Postfixation

The hybridizations are performed in 10-mL Reacti-Vials (*see* Note 1) or screw-cap plastic tubes. Blastocysts are best handled by serial transfer to glass dishes containing the appropriate solutions. It is best to perform experiments on groups of 10 or more embryos, both to provide for losses and simplify interpretation of the results.

1. Rehydrate the embryos on ice through 75, 50, and 25% methanol:PBS (5 min each).
2. Wash the embryos three times at room temperature with PBT (5 min each).
3. Wash the embryos three times for 30 min room temperature with RIPA.
4. Fix the embryos for 20 min at room temperature in 4% paraformaldehyde, 0.2% glutaraldehyde in PBT with occasional mixing (*see* Note 6).
5. Wash the embryos three times for 5 min at room temperature with PBT.

3.3. Prehybridization and Hybridization

1. If the pretreatments have been performed in a batch, separate the groups of embryos to be hybridized with different probes. Set up groups of embryos for the three controls, one for a heterologous probe, one for no probe, and one for no probe and no antibody.
2. Wash the embryos at room temperature with 1:1 hybridization buffer:PBT. All the embryos should sink within 5–10 min.
3. Wash the embryos once with hybridization buffer.
4. Prehybridize the embryos at 70°C, 1–5 h in about 1 mL of hybridization buffer containing 100 µg/mL tRNA and 100 µg/mL sheared, denatured herring sperm DNA.
5. Remove the prehybridization solution and add 0.5–1.0 mL of hybridization solution plus tRNA and DNA containing a 1:100–1:200 dilution of freshly denatured digoxigenin-labeled RNA probe (1.0–1.5 ng/µL). If using the Pierce Reacti-vials, place a foam rubber insert containing an open 0.5-mL Eppendorf tube filled with hybridization buffer into the neck of the vial (but well above embryos) to humidify the chamber.
6. Coat the rim of the vial with vacuum grease and seal firmly (make sure the Teflon side of the cap is facing down). Mix gently but thoroughly (avoiding mixing in the contents of the humidifying tube). Incubate at 70°C overnight in a water bath or on a heating block (*see* Note 7).

3.4. Posthybridization Washes and Antibody Conjugate Binding

1. Wash once or twice for 10 min at 70°C with hybridization buffer.
2. At this point begin preadsorbing the antidigoxigenin Fab/alkaline phosphatase conjugate (*see* Section 2.).

3. Wash twice for 5 min at 65°C with 2X SSC pH 4.5, 50% formamide, 0.1% Tween-20.
4. Wash three times for 30 min with the same solution at 65°C.
5. Allow the embryos to cool to room temperature.
6. Wash three times at room temperature with 1X TBST.
7. Block the embryos by incubating for 1 h in 10% heat-inactivated sheep serum in TBST at room temperature.
8. Replace the sheep serum with preadsorbed, diluted antibody conjugate.
9. Mix the antibody and embryos well. Incubate the samples overnight at 4°C (*see* Note 8). During this step the antibody will bind to probe captured by the target mRNA.

3.5. Post-Antibody Conjugate Washes and Staining

1. Remove the antibody conjugate solution and wash three times for 5 min with TBST at room temperature.
2. Wash for three to five times, 30–60 min each wash, with TBST at room temperature.
3. Wash three times for 10 min with alkaline phosphatase buffer. During this step the pH is raised to the optimum for alkaline phosphatase activity.
4. Transfer the embryos to a shallow glass dish for staining (*see* Note 9).
5. Add 1–2 mL of freshly prepared staining solution. Swirl well, then cover the dishes and place them in the dark.
6. The embryos can be removed from the dark briefly for observation under a semidark-field dissecting microscope. If the target mRNA is detected in a tissue then a dark purple color stain will appear. Signals corresponding to abundant RNA species (e.g., actin) should be visible within 15 min, less abundant species (e.g., Krox-20) in 2 h (*see* Note 10).
7. Stop the staining reaction by rinsing the embryos three times in PBT containing 1 m*M* EDTA. The stained patch on the embryo is stable for at least several weeks in PBT at 4°C (*see* Note 11).

4. Notes

1. The Reacti-Vials have steep conical bottoms, making them ideal for both large volume washes and smaller volume incubations. They also optically magnify the embryos. Embryos sink to the bottom of the vials in all buffers used. During washes, always leave the embryos in a small amount of fluid. Never allow them to become dry, as they will either disintegrate or display high background. Be careful when performing the washes as it is very easy to accidently lose embryos.

2. Antisense riboprobes are synthesized as runoffs from linearized plasmid templates using bacteriophage RNA polymerases (T3, T7, SP6) (*see* Chapter 12). Digoxigenin UTP is substituted for UTP at a 1:4 ratio. Plasmids isolated by the Qiagen method and phenol extracted after restriction digests work well. It is not necessary to undertake CsCl-EtBr preps of plasmid. Probes ranging from 400–1500 bp have been used, but longer probes may also be suitable. It does not seem necessary to hydrolyze the riboprobe before use. Probe concentrations are in the hundreds of nanograms-microgram per milliliter range, much higher than those typically employed in radioactive protocols.

3. Normally 6- to 10-d-old embryos are used, but other stages, e.g., blastocysts, have been processed successfully (*see also* Note 11)

4. Dehydration in methanol is important for optimal signal strength.

5. Embryos have been stored with satisfactory results for up to 3 wk, but longer times may be suitable as well.

6. Fixation is a critical step in the procedure; overfixing will diminish signal, underfixing will result in the disintegration of the embryos during subsequent steps.

7. The unusually high stringency and low pH of the hybridization conditions are based on empirical observations from the *Drosophila* system. The high temperature (and possibly also the acid conditions) is probably important for the inactivation of endogenous alkaline phosphatase activity.

8. The long incubation time increases sensitivity, but more abundant RNA species can be visualized with shorter incubations of several hours.

9. Glass dishes are used for staining as plastic may induce a precipitate.

10. Incubation in staining solution can be continued as long as overnight but background may begin to be a problem, especially in embryonic or extraembryonic cavities.

11. Staining in the control without added probe or antibody indicates that endogenous alkaline phosphatase has been detected (a problem we have never experienced). However, a problem that is more commonly encountered is background caused by trapping of antibody (staining visible in control without added probe) and/or probe (staining visible in control without probe, and also in control with heterologous probe) in embryonic cavities, such as the amniotic cavity, heart, or brain ventricles. Background problems become particularly acute in embryos older than 9 d. Such difficulties can be alleviated by dissecting open cavities before hybridization or performing the procedure on the isolated organ/region of interest.

References

1. Rosen, B. and Beddington, R. S. P. (1993) Whole-mount *in situ* hybridization in the mouse embryo: gene expression in three dimensions. *TIG* **9,** 162–167.
2. Tautz, D. and Pfeile, C. (1989) A non-radioactive in situ hybridization method for the localization of specific RNA's in *Drosophila* embryos reveals translational control of the segmentation gene *hunchback*. *Chromosoma* **98,** 81–85.

CHAPTER 31

PACE (Probe Assay— Chemiluminescence Enhanced)

A Magnetic Bead Assay for the Noncultural Diagnosis of Gonorrhea

Paul A. Granato

1. Introduction

Noncultural methods are available for the direct detection of various microorganisms in clinical specimens and/or for the identification of certain microbial agents recovered from culture. The Gen-Probe PACE 2 (Probe Assay—Chemiluminescence Enhanced) system for *Neisseria gonorrhoeae* uses the noncultural technique of nucleic acid hybridization *(1)* to detect *Neisseria gonorrhoeae* directly in urogenital swab specimens and/or to identify isolates of *N. gonorrhoeae* when recovered from culture. Clinical evaluations *(2,3)* of this commercially available PACE 2 assay (Gen-Probe, San Diego, CA) have found it to be at least comparable to conventional cultural technologies for the detection and/or characterization of gonococci.

Nucleic acid hybridization tests are based on the ability of complementary nucleic acid strands to specifically align and associate to form stable double-stranded complexes *(1)*. The Gen-Probe PACE 2 system uses an acridinium labeled, single-stranded DNA probe that is complementary to the ribosomal RNA of *N. gonorrhoeae*. After the ribosomal RNA is released from the test organism, the labeled

From: *Methods in Molecular Biology, Vol. 28:*
Protocols for Nucleic Acid Analysis by Nonradioactive Probes
Edited by: P. G. Isaac Copyright ©1994 Humana Press Inc., Totowa, NJ

DNA probe combines with the target organism's ribosomal RNA to form a stable DNA:RNA hybrid. Hybridized labeled probe is separated from the nonhybridized labeled probe by the addition of a suspension of magnetic particles that specifically bind to hybridized probe. The magnetic particles are washed and the hybridized probe is eluted from the magnetic particles by the addition of an elution reagent. The supernatant fluid is transferred to a separate tube and assayed for the amount of hybridized probe present in the sample by the addition of a detection reagent. This reagent hydrolyzes esterified acridinium from the DNA probe resulting in the production of light. The amount of light generated from this chemiluminescent reaction is measured quantitatively in the Gen-Probe Leader luminometer and is directly proportional to the amount of target RNA originally present in the test sample.

Interpretations of the assay are based on the ratio of the response (in relative light units) of the specimen and the mean of three negative controls. The Leader luminometer calculates the results automatically and prints a hard-copy result for each test sample. This Gen-Probe chemiluminescence assay may be used to detect the presence of *N. gonorrhoeae* directly in urogenital samples or to confirm the identity of possible gonococcal isolates.

2. Materials

The Gen-Probe PACE 2 system for *Neisseria gonorrhoeae* is commercially available as a kit from Gen-Probe, San Diego, CA, and includes the following reagents:

1. STD probe diluent: Buffered solution (*see* Note 1).
2. Probe reagent: This is supplied in a lyophilized form, and needs to be reconstituted before use as follows. Remove the probe diluent from the kit and vortex for 10 s. Warm the probe diluent by swirling the vial in a water bath at 60°C for 3–4 min. Vortex again for 10 s to ensure a homogeneous solution. Pipet 6.0 mL of probe diluent into lyophilized probe reagent. Allow the reagent to stand at room temperature for 2 min and then vortex for 10 s prior to use. Visually inspect to ensure that the reagent is completely rehydrated and homogeneous. Record on the label the date reconstituted. The reconstituted probe reagent is stable for 3 wk when stored at 2–8°C or until the date stamped on the reagent container, whichever is first (*see* Note 1).
3. STD activator: Buffered solution.

4. STD separation reagent: Solid phase in a buffered solution containing 0.02% sodium azide. Store at 2–8°C, do not freeze.
5. STD wash solution: Buffered solution.
6. Positive control: Noninfectious *N. gonorrhoeae* nucleic acid in a buffered solution.
7. STD negative reference: Noninfectious nucleic acid in a buffered solution.
8. Gen-Probe detection reaction kit (provided separately):
   a. Detection reagent I: 0.1% hydrogen peroxide in $0.001N$ nitric acid.
   b. Detection reagent II: $1N$ sodium hydroxide.
9. Separation solution: First determine the number of tests to be performed, then calculate the volumes of activator and separation reagents as follows:

With Eppendorf repeating pipetor:

volume of activator (mL) = number of tests + 2 extra tests

With with bottle top dispenser:

volume of activator = number of tests + 10 extra tests
volume of seperation reagent = volume of activator/20

Pour the required volume of activator into a clean, dry container. Mix the separation reagent, add the required volume to the activator, and mix well. The prepared separation solution is stable for 6 h after preparation when stored at 20–25°C.

Other reagents contained in the Gen-Probe kit should be stored at 2–25°C and are stable until the date stamped on the container.

3. Methods

3.1. Sample Collection and Preparation

The Gen-Probe PACE 2 system for *Neisseria gonorrhoeae* is designed to detect the presence of *N. gonorrhoeae* in specimens obtained from the male urethra and the female endocervical canal using the Gen-Probe PACE specimen collection kit (*see* Note 2). It may also be used to identify *N. gonorrhoeae* isolated from culture (*see* Note 3). This collection kit contains two glass beads and 1.0 mL of preservative that consists of a bacterial lysing detergent with nuclease inhibitors to prevent hydrolysis of ribosomal RNA. Only swabs contained in the PACE specimen collection kit can be used to collect patient specimens, and swabs collected from patients or colonies isolated from culture **must** be transported to the laboratory in the Gen-Probe transport medium. Collect swab samples as follows.

3.1.1. Cervical Swab Specimens

1. Remove excess mucus from the cervical os and surrounding mucosa using one of the swabs provided in the cervical collection kit and discard the swab.
2. Insert the second swab from the collection kit into the endocervical canal.
3. Rotate the swab for 10–30 s in the endocervical canal to ensure adequate sampling.
4. Withdraw the swab carefully; avoid any contact with the vaginal mucosa.
5. Insert the swab into the Gen-Probe transport tube.
6. Break off or cut the swab shaft to fit the tube and cap the tube.

3.1.2. Urethral Swab Specimens

1. Patient should not have urinated for at least 1 h prior to sample collection.
2. Collect the urethral exudate or insert the swab from the urethral collection kit 2–4 cm into the urethra using a rotating motion to facilitate insertion.
3. Once inserted, rotate the swab gently using sufficient pressure to ensure the swab comes into contact with all urethral surfaces. Allow the swab to remain inserted for 2–3 s.
4. Withdraw the swab.
5. Insert the swab into the Gen-Probe transport tube.
6. Cut the swab shaft to fit the tube and cap the tube.

3.1.3. Culture Isolates
from Modified Thayer-Martin or Chocolate Agar Plates

1. Using an inoculating loop, remove a sufficient number of colonies from a culture plate that has been properly incubated for no longer than 24 h. Prepare a bacterial suspension equal to a #1 MacFarland standard (*see* Note 4) in sterile saline and vortex.
2. Pipet 100 μL of the prepared suspension into the Gen-Probe transport tube and vortex.

3.2. Storage of Specimens

Transport the tubes to the laboratory at 2–25°C and store at 2–25°C until tested. Samples should be assayed with the Gen-Probe PACE 2 system within 7 d. If longer storage is necessary, process the specimen as described in Section 3.1. and freeze at –20 to –70°C.

3.3. Test Procedure

The PACE 2 assay for *N. gonorrhoeae* consists of four basic steps: sample preparation, hybridization, separation of hybridized from unhybridized probe, and, lastly, activation and chemiluminescent measurement. The probe assay takes about 2 h to complete and positive and negative control samples must be included in each test analysis (*see also* Note 5). For convenience, repeating pipetors or dispensers may be used for addition of probe solution, separation solution, and wash solution. Pipetors with disposable tips are recommended for pipeting specimens and controls to avoid sample carryover and crosscontamination. Care should be taken to pipet probe reagent to the bottom of tubes without inserting the pipet tip into the tubes or touching the tip to the rim of each tube.

1. Sample preparation: Allow the specimens to reach room temperature prior to processing.
2. Vortex each Gen-Probe transport tube for at least 5 s. Express all liquid from the swab by pressing the swab against the wall of the tube. Discard the swab.
3. Label tubes with sample identification numbers. Include three tubes for the negative reference and one for the positive control.
4. Insert the tubes into the tube rack of the magnetic separation unit.
5. Vortex each specimen for 5 s.
6. Pipet 100 µL of each of the controls and specimens to the bottom of the respective tubes (*see* Note 6).
7. Pipet 100 µL of the probe reagent to the bottom of each tube, taking care not to touch the top or sides of the tube.
8. Cover the tubes with adhesive sealing cards ensuring that each tube is sealed.
9. Shake the rack three to five times to mix.
10. Incubate the tubes in a water bath at $60 \pm 1°C$ for 1 h (*see* Note 7). Hybridization occurs during this step.
11. Prepare the Gen-Probe Leader for operation. Make sure there is sufficient volume of detection reagents I and II to complete the tests (200 µL of each are required per sample).
12. Remove the tube rack from the water bath and remove the sealing cards.
13. Pipet 1 mL of the well mixed, prepared separation solution into each tube, and mix well.
14. Cover the tubes with sealing cards and shake the tube rack three to five times to mix.
15. Incubate the tubes in a water bath at $60 \pm 1°C$ for 10 min.
16. Remove the tube rack from the water bath. Remove the sealing cards and place the tube rack on the base of the Gen-Probe magnetic separation unit for 5 min at room temperature.

17. Holding the tube rack and base of the magnetic separation unit together, decant the supernatants. Before turning the tubes upright, shake the unit two or three times and then blot the tubes on absorbent paper (*see* Note 8).
18. **Do not remove the tube rack from the Gen-Probe magnetic separation base.** Fill each tube to the rim with wash solution.
19. Allow the tubes to remain on the magnetic separation base for 20 min at room temperature.
20. Holding the tube rack and base together, decant supernatants. **Do not blot.** Approximately 50–100 µL of wash solution should remain in each tube.
21. Separate the tube rack from the base and shake the tube rack to resuspend the pellets.
22. Select the appropriate protocol from the Leader software.
23. Using a damp tissue or damp paper towel, wipe each tube to ensure that no residue is present on the outside of the tube. Ensure that the pellets are resuspended and insert the tubes in Leader according to the prompts provided by the instrument software (*see* Note 9). Read the tubes in the following order:
    a. Negative reference, 3 tubes
    b. Positive control, 1 tube
    c. Specimen tubes
24. When the analysis is complete, remove the tube(s) from the Leader.

3.4. Calculation of Results

1. The results of the Gen-Probe PACE 2 system for *Neisseria gonorrhoeae* are calculated based on the difference between the response in relative light units (RLU) of the specimen and the mean of the negative reference.

   Mean of the negative reference = sum of the three negative reference replicates divided by 3

   Example:

   Mean of the negative reference = (55 RLU + 60 RLU + 50 RLU)/3 = 55

   Specimen response = 894 RLU

   Difference = 894 RLU – 55 RLU
   = 839 RLU Positive

2. The Leader prints the specimen response and compares this response to an assigned assay cutoff. A positive or negative interpretation as compared to this cutoff is printed.

3.5. Interpretation of Results (see *Note 10*)

1. Direct specimen:
   a. Positive—The difference is ≥300 RLU. A positive result indicates that *Neisseria gonorrhoeae* is present in the specimen tested and strongly supports a diagnosis of gonococcal infection.

  b. Negative—The difference is <300 RLU. A negative result indicates the absence of *Neisseria gonorrhoeae* in the specimen tested (*see* Note 11).
2. Culture confirmation (*see* Note 12)
  a. Positive—The difference is ≥10,000 RLU.
  b. Negative—The difference is <10,000 RLU.

3.6. Quality Control and Acceptability of Results

1. Negative reference:
  a. The response of each negative reference should be <200 RLU and >20 RLU.
  b. All negative reference values should fall within 30% of the mean response for the negative reference.
2. Positive control: The difference in the response of the positive control and the mean response of the negative reference should be >600 RLU.

4. Notes

1. Probe reagent: Heating and swirling of the probe diluent and reconstituted probe solution at 60°C is imperative to prevent gel formation and ensure a homogenous solution. The reconstituted probe reagent is stable for 3 wk when stored at 2–8°C or until the date stamped on the reagent container, whichever is first. If the reconstituted probe reagent has been refrigerated, vortex for 10 s, then warm it by swirling the vial in a waterbath at 60°C for 2 min. Prior to use, vortex again for 10 s to ensure homogeneity. It may be necessary to repeat this procedure if the reconstituted probe reagent is not homogeneous.
2. During routine analysis, bloody specimens have not proven to interfere with assay performance. However, grossly bloody specimens (>80 µL whole blood in 1 mL transport media) may interfere with performance.
3. The Gen-Probe PACE 2 assay has been tested using urogenital specimens and cultural isolates only. Performance with other clinical specimens (i.e., pharyngeal and/or rectal) has not been assessed. As such, this assay is not intended for use in detecting *N. gonorrhoeae* in specimens collected from extragenitourinary sites.
4. A #1 MacFarland standard is a suspension of a fine white precipitate of barium sulfate, which, when shaken up, approximates the turbidity of a bacterial suspension containing 3.0×10^8 organisms/mL. A #1 MacFarland standard is prepared by adding 0.1 mL of a 1% solution of barium chloride to 9.9 mL of 1% sulfuric acid.
5. As in any reagent system, excess powder on some gloves may cause contamination of opened reagents or reaction tubes. Gen-Probe recom-

mends that customers experiencing difficulty with the test avoid using this type of laboratory glove. Use of powderless gloves (no talcum powder) will avoid this difficulty.

6. Occasionally a specimen may be too viscous to pipet. Be sure that specimens are at room temperature and vortex to liquify.

7. Temperature: The hybridization and separation reactions are temperature dependent. Therefore, it is imperative that the water bath and reaction tubes be equilibrated uniformly during these steps. A covered water bath capable of maintaining $60 \pm 1°C$ should be used.

8. Blotting: Discard absorbent paper after each blotting to avoid contamination. **Do not blot after the wash step.**

9. Detection: Tubes should be read on the Leader within 60 min of decanting the wash solution. Tubes should be maintained at 20–25°C prior to reading.

10. As in any clinical situation, diagnosis should not be based on the results of a single laboratory test. If the test result is negative and the clinical indications strongly suggest gonococcal infection, additional specimens should be collected for further testing.

11. A negative test result does not exclude the possibility of gonococcal infection because test results may be affected by improper specimen collection. Results from the Gen-Probe PACE 2 assay should be interpreted in conjunction with other laboratory and clinical data.

12. When using the culture confirmation procedure, the organism must be viable to obtain a valid result.

References

1. Kohne, D. E., Steigerwalt, A. G., and Brenner, D. J. (1984) Nucleic acid probe specific for members of the genus Legionella, in *Legionella: Proceedings of the 2nd International Symposium* (Thronsberry, C., Balows, A., Feeley, J. C., and Jakubowsky, W., eds.), American Society for Microbiology, Washington, DC, pp. 107,108.
2. Granato, P. A. and Franz, M. R. (1990) Use of the Gen-Probe PACE system for the detection of *Neisseria gonorrhoeae* in urogenital samples. *Diag. Microbiol. Infect. Dis.* **13,** 217–221.
3. Granato, P. A. and Franz, M. R. (1989) Evaluation of a prototype DNA probe test for the noncultural diagnosis of gonorrhea. *J. Clin. Microbiol.* **27,** 632–635.

CHAPTER 32

Detection of Foodborne Pathogens Using DNA Probes and a Dipstick Format

E. Patrick Groody

1. Introduction

The contamination of food with organisms, such as *Salmonella*, *Listeria*, and *Escherichia coli*, continues to be an area of concern for consumers and food producers. Historically, the identification of such organisms has been done using conventional culture methods and biochemical techniques. These methods, although still considered by most to be the "gold standard" for microbial identification, can often be time consuming and laborious to perform. In addition, these methods can also be complicated by the heterogeneous nature of food microflora and the subjective nature of many microbiological and biochemical techniques.

During recent years, many new methods have been developed for detecting foodborne pathogens. These methods no longer rely on conventional microbiological and biochemical techniques. Instead, DNA probes or immunological reagents are used to screen for the presence of the organism of interest. Such new methods have helped improve the objectivity of test methods as well as reduce the time required to test many food products.

DNA probe based methods have become increasingly popular in recent years. These methods typically employ short nucleic acid fragments to detect an organism's unique nucleic acid sequence. Such

From: *Methods in Molecular Biology, Vol. 28:*
Protocols for Nucleic Acid Analysis by Nonradioactive Probes
Edited by: P. G. Isaac Copyright ©1994 Humana Press Inc., Totowa, NJ

methods offer a variety of advantages, including sensitive nonradio-active detection and elimination of the need to isolate the target organism. In addition, by carefully selecting the nucleic acid target (i.e., RNA or DNA , and so on) and the probe sequences, the specificity of the technique can be adjusted to fit the needs of particular applications. For example, when the detection of a single species is desired (i.e., *Staphylococcus aureus*), probes may be designed such that all strains of *S. aureus* can be detected without detecting other organisms belonging to the genus *Staphylococcus*. Alternatively, when broader inclusivity is desired, as in the case of a screening test for *Salmonella*, sets of probes that detect many different *Salmonella* species can be employed.

The overall format of DNA probe tests generally involve five basic steps. In most cases, cultural enrichment of the test sample is required in order to increase the number of organisms to detectable levels. Once the cultural enrichment is complete, the assay is then completed by lysing the target organisms, hybridizing the probes to the targets, immobilizing the probe target complex onto a solid support and detecting the immobilized probe-target complex.

In order to develop a DNA probe test, an extensive understanding is required of the polynucleotide sequences of the intended target organism, genetically related organisms, and other organisms that might be encountered in the test sample. Generally, sequence information for such organisms can be found in various databases, however, often it is necessary to supplement this information with additional laboratory experiments. Once the nucleic acid sequences of the target organisms are determined, probes that are complementary to these sequences are designed and then produced by a variety of chemical or biological methods. Candidate probes are then tested using extensive panels of organisms in order to demonstrate that the desired specificity properties have been achieved. If necessary, probe specificity can be refined during the assay development process by carefully controlling the time of incubation, and the pH, salt concentration, and temperature of the medium used during incubation of the probes with the test sample.

Numerous DNA probe assays for the detection of foodborne pathogens, such as *Listeria (1)*, *Salmonella (2,3)*, and *E. coli (4)*, have been developed. Initially, most such methods had limited utility owing to the use of isotopic signal generating systems and complex assay formats. More recently, however, a variety of format improvements and the development of more sensitive nonradioactive detection methods have helped to broaden the application of probe based methods *(5,6)*. A number of commercially available kits for detecting *Salmonella*, *E. coli*, *Listeria*, *S. aureus*, *Campylobacter*, and *Yersinia* are now available from companies such as GENE-TRAK systems (Framingham, MA). These products have greatly enhanced the utility of probes for use in routine food testing in commercial testing laboratories.

A schematic representation of a typical DNA probe test for the detection of *Salmonella* in food is shown in Fig. 1. In this test, a dipstick format is employed. The probes used consist of a mixture of capture probes and detector probes. The capture probes contain two specific binding regions. The first region contains a *Salmonella*-specific nucleic acid sequence. The second region contains a polydeoxyadenylate tail. This serves to link the probe target complex to a polydeoxythymidylate coated solid support. Similarly, the detector probes also contain *Salmonella*-specific sequences. These sequences are labeled with fluorescein groups that serve to bind an antibody-enzyme conjugate to the immobilized probe-target complex. After removal of all of the excess reactants and cellular debris, the plastic dipstick containing the immobilized probe-target complex is added to a substrate-chromogen mixture in order to generate a highly colored product when *Salmonella* is present in the original test sample.

2. Materials

All materials required for detection of *Salmonella* in food products can be purchased in kit form (Colorimetric GENE-TRAK *Salmonella* Assay) from GENE-TRAK systems.

1. Lysis solution: $0.75M$ NaOH.
2. Neutralization solution: $2M$ Tris-HCl, pH 7.5.
3. *Salmonella* probe solution: 1–2 µg/mL capture probe, 1–5 µg/mL detector probe in $1.05M$ Tris-HCl, pH 7.5, 0.5 mM EDTA, 0.05% BSA, 0.005% NP-40. Probe solution should be stored at 2–8°C.
4. Wash solution: 50 mM Tris-HCl, pH 7.5, 2 mM EDTA, 100 mM NaCl, 0.1% Tween-20.
5. Enzyme conjugate: Antifluorescein antibody conjugated to horseradish peroxidase. Conjugate should be stored at 2–8°C.

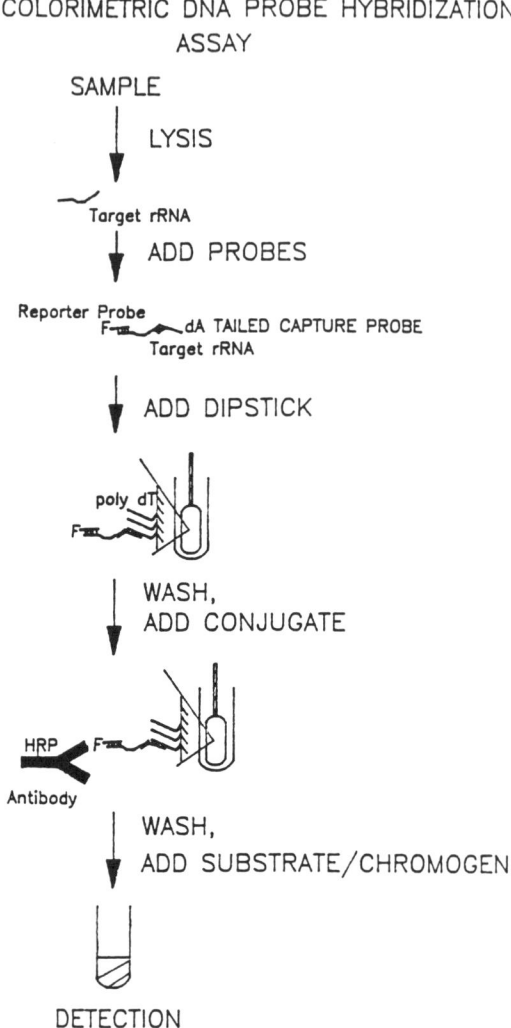

Fig. 1. A schematic representation of the steps required to detect *Salmonella* using the DNA probe test.

6. Diluted enzyme conjugate solution: This should be prepared immediately before use (during step 10 of the assay procedure). Dilute the enzyme conjugate concentrate 100-fold with wash solution. Do not store the diluted enzyme conjugate for greater than 60 min.
7. Substrate solution: Urea peroxide in aqueous buffer. Store at 2–8°C in a dark container. Do not expose to light.

8. Chromogen solution: Tetramethylbenzidine in organic solution. Store at 2–8°C.

9. Substrate/chromogen reagent: This is prepared immediately before use (step 14 of the assay procedure). Add 2 vol of substrate solution to 1 vol of chromogen solution (0.75 mL is needed for each sample). Do not store this reagent for longer than 60 min.

10. Stop solution: 2*M* sulfuric acid. **Caution:** Solution is corrosive and should be handled with extreme care. Spills should be wiped up immediately. Do not allow to come in contact with skin. Wash all exposed areas immediately with copious amounts of water.

11. Dipsticks (poly-dT coated): Store at room temperature in a dessicated container.

12. Positive control (suspension of killed *Salmonella typhimurium* sufficient to give an OD at 450 nm equal to 1.0 or greater.). Store at 2–8°C.

13. Negative control (suspension of killed *Citrobacter freundii,* concentration sufficient to give an OD at 450 nm <0.1). Store at 2–8°C.

14. Lactose broth: from Difco Laboratories (Detroit, MI) *(7) (see* Note 1).

15. Tetrathionate (TT) broth: from Difco Laboratories *(7) (see* Note 1).

16. Selenite cysteine (SC) broth: from Difco Laboratories *(7) (see* Note 1).

17. Gram negative (GN) broth: from Difco Laboratories *(7) (see* Note 1).

18. A supply of 12 × 75-mm glass test tubes.

19. Photometer capable of reading the OD at 450 nm of a solution in 12 × 75-mm tubes (available from GENE-TRAK systems).

3. Methods
3.1. Cultural Enrichment of Samples

This section describes a protocol for bacterial enrichment of raw meat and raw milk samples. Variations of this procedure for use with other foods are described in the Notes.

1. Homogenize 1 vol of the food sample in 9 vol of lactose broth. For enrichment of other samples, *see* Note 2.

2. Incubate the mixture for 22–24 h at 35°C (preenrichment culture).

3. Remove the preenrichment culture from the incubator and mix well.

4. Transfer 1 mL of preenrichment culture to 10 mL of TT broth. Transfer a second 1-mL aliquot to 10 mL of SC broth. Both broths are required to ensure that acceptable titers of various *Salmonella* can be obtained. Failure to use both broths may lead to inaccurate results with some strains of *Salmonella*.

5. Incubate the TT and SC broths for 16–18 h at 35°C (for other foods, *see* Note 3).

6. Remove the TT and SC cultures from the incubator and transfer 1 mL of each culture to separate tubes, each containing 10 mL of GN broth in order to increase the titer of *Salmonella* (if present) to detectable levels. Although, in some cases, *Salmonella* can be detected directly for the TT and SC broths, broad screening of many different food types is most accurately conducted by using the complete enrichment scheme. Failure to do so could lead to an increase in the incidence of false negative results.

7. Incubate the GN cultures for 6 h at 35°C (for other foods, *see* Note 4) and then remove 0.25 mL of each GN culture to perform the probe assay.

3.2. Assay Procedure

1. Equilibrate a water bath at 65°C (*see* Note 5).
2. Prepare four wash basins each containing 300 mL of wash solution. Place one wash basin in the 65°C water bath. The remaining three wash basins should be left at room temperature. Allow all of the reagents to warm to room temperature before use (*see* Note 5).
3. Mix all of the GN cultures thoroughly before use.
4. For each sample to be tested, add 0.25 mL of each respective GN culture (one from TT and one from SC for each food sample) to a 12 × 75-mm glass tube. Total test sample volume should be 0.5 mL/food sample.
5. Set up the positive and negative controls. Mix the positive control and negative control reagents thoroughly before use. Add 0.5 mL of positive control and 0.5 mL of negative control to separate 12 × 75-mm glass tubes.
6. Add 0.1 mL of lysis solution to each tube. Shake the tubes thoroughly and incubate at room temperature for 5 min.
7. Add 0.1 mL of neutralization solution to each tube and shake all the tubes thoroughly. Cover the tubes and place them in the 65°C water bath. Incubate at 65°C for 15 min.
8. When the 15-min incubation is complete, add 0.1 mL of probe solution to each tube. Cover the tubes and incubate for 15 min at 65°C.
9. When the 15-min incubation is complete, remove the cover from the tubes and add a plastic poly-dT coated dipstick to each tube. Incubate the dipsticks in the tubes for 1 h at 65°C.
10. During the 1-h incubation step, dilute the enzyme conjugate solution (*see* Section 2.). Do not store the diluted solution for >60 min.
11. For each sample to be tested, add 0.75 mL of diluted enzyme conjugate solution to a separate 12 × 75-mm glass test tube.

12. After incubation of the dipsticks with the test samples for 1 h, remove the dipsticks from the tubes and place them in the wash basin at 65°C; wash them with gentle shaking for 1 min. Remove the dipsticks from the wash basin and place them in one of the wash basins at room temperature. Wash them with gentle shaking at room temperature for 1 min.

13. Blot the dipsticks on absorbent paper and then add them to the tubes containing the diluted enzyme conjugate. Incubate the dipsticks in enzyme conjugate at room temperature for 20 min.

14. During the 20-min incubation, prepare enough substrate/chromogen reagent for each sample being tested (*see* Section 2.). Do not store the combined reagents for >60 min.

15. Add 0.75 mL of substrate/chromogen mixture to a separate 12 × 75-mm glass tube for each sample being tested.

16. When the incubation of the dipstick with the enzyme conjugate is complete, remove the dipsticks from the tubes and wash them sequentially in the remaining two wash basins for 1 min each. Do not wash in the wash solutions used in the previous steps.

17. Blot the dipsticks on absorbent paper and place them in the tubes containing the substrate-chromogen mixture. Incubate the dipsticks in the tubes for 20 min at room temperature.

18. Remove the dipsticks from the tubes and discard them.

19. Add 0.25 mL of stop solution to each tube containing the substrate-chromogen mixture. Shake the tubes thoroughly to make sure all the reagents are completely mixed.

20. Determine the OD of each solution at 450 nm. Samples are considered positive if the absorbance reading exceeds the OD of the negative control by at least 0.1 OD unit. The OD of the negative control must be ≤0.1 at 450 nm, and the OD of the positive control should be >1.0 at 450 nm (*see* Note 5). Positive samples should be confirmed by appropriate biochemical confirmation tests *(7)*.

4. Notes

1. Alternative cultural enrichment media may be used but are not recommended by the kit manufacturer.

2. Other foods can be enriched as described in BAM/AOAC guidelines *(7)*.

3. Samples from other foods should be incubated for 6 h.

4. Samples from other foods should be incubated for 12–18 h.

5. Care should be taken to ensure that all reagents are maintained at the appropriate temperature. In particular, the water bath must be maintained at 65 ± 1°C for proper results. Failure to do so may result in high back-

grounds and false positive results if the temperature is too low, or low signal and false negative results if the temperature is too high.

References

1. King, W., Raposa, S., Warshaw, J., Johnson, A., Halbert, D., and Klinger, J. D. (1989) A new colorimetric DNA hybridization assay for *Listeria* in foods. *Int. J. Microbiol.* **8,** 225–232.

2. Fitts, R., Diamond, M., Hamilton, C., and Neri, M. (1983) DNA-DNA hybridization assay for detection of *Salmonella spp.* in foods. *Appl. Environ. Microbiol.* **46,** 1146–1151.

3. Flowers, R. S., Klatt, M. J., Mozola, M. A., Curiale, M. S., Gabis, D. A., and Silliker, J. H. (1987) DNA hybridization assay for the detection of *Salmonella* in foods. *J. Assoc. Off. Anal. Chem.* **70,** 521–529.

4. Hill, W. E. (1981) DNA hybridization method for detecting enteropathogenic *Escherichia coli* in human isolates and its possible application to food samples. *J. Food Safety* **3,** 233–247.

5. Hsu, H. Y., Chan, S. W., Sobel, D. I., Halbert, D. N., and Groody, E. P. (1991) A colorimetric DNA hybridization method for the detection of *Escherichia coli* in foods. *J. Food Prot.* **54,** 249–255.

6. Wilson, S. G., Chan, S., Deroo, M., Vera-Garcia, M., Johnson, A., Lane, D., and Halbert, D. N. (1990) Development of a colorimetric second generation nucleic acid hybridization method for the detection of *Salmonella* in foods and a comparison with conventional culture procedure. *J. Food Prot.* **55,** 1394–1398.

7. U.S. Food and Drug Administration (1991) *Bacteriological Manual*, 7th ed. Association of Official Analytical Chemists, Arlington, VA.

CHAPTER 33

Analysis of Gene Sequences by Hybridization of PCR-Amplified DNA to Covalently Bound Oligonucleotide Probes

The Reverse Dot Blot Method

Ernest S. Kawasaki and Farid F. Chehab

1. Introduction

The introduction of polymerase chain reaction (PCR) technologies *(1–4)* has greatly simplified the analysis of gene sequences. Prior to PCR, comparison of genes and their alleles required lengthy, tedious, and somewhat difficult procedures. These steps included DNA or RNA purification, cloning of DNA restriction fragments or complementary DNA, subcloning, isolation of the corresponding plasmid, and sequencing of the inserts. PCR has eliminated much of the tedium by allowing amplification of a desired sequence from small amounts of unpurified nucleic acids and direct sequencing of the amplified product. In certain cases one does not need to know the entire sequence of a gene, but only requires detection of a specific allele or mutation. This is true in instances where the sequences of the the majority of the alleles or mutations in the population are already known. For example, the RAS oncogene system *(5)* has three loci with cancer-associated mutations occurring at three codons, 12, 13, and 61. Over 60 mutations have been discovered at the cystic fibrosis (CF) locus

From: *Methods in Molecular Biology, Vol. 28:*
Protocols for Nucleic Acid Analysis by Nonradioactive Probes
Edited by: P. G. Isaac Copyright ©1994 Humana Press Inc., Totowa, NJ

(6), but a few predominant mutations in the caucasian population account for 70–90% of the known cases. Similarly, the HLA Class II locus is very polymorphic *(7)*, and detection of allelic, single bp differences are important in forensics, paternity determination, transplantation, and so on.

If only a few samples are to be analyzed, direct sequencing of PCR products is a feasible alternative to former cloning methods. However, studies of cancer-associated gene mutations, genetic diseases, and HLA polymorphisms usually involve large numbers of samples and relying solely on sequencing would be too cumbersome and time consuming. As alluded to above, if the mutations or allelic differences are already known, one only needs to identify bp changes at specific codons of the gene in question. To do this, methods have been developed that rely on hybridization of allele specific oligonucleotides (ASOs) to PCR products that are "dot blotted" onto nitrocellulose or nylon membranes *(8)*. In this protocol the amplified DNAs are bound to the membranes, and then annealed with short oligonucleotide probes that are able to discriminate between nucleic acid sequences that differ by only one nucleotide *(9)*. Although nonamplified DNAs can be used with this method, the complexity of the mammalian/human genome $(3 \times 10^9 \text{ bp})$ makes background hybridization "noise" a formidable problem. Amplification of DNA prior to ASO hybridization has eliminated much of this difficulty by increasing the concentration of the target sequence many million-fold. In this way, PCR amplification of the desired sequence in conjunction with the dot blot method and ASO hybridization has made the analysis of mutations or polymorphisms in the human genome much easier.

Even though the dot blot method is well suited to handling large numbers of samples, it can become unwieldy when the screening process requires a large number of probes. For example, with CF a thorough analysis of the DNA from a single individual would involve more than 60 different oligonucleotides probings, and a similar situation exists for the RAS and HLA systems. To overcome this problem, a modification of the dot blot protocol was developed called the "reverse dot blot" *(10)*. In this case, the oligonucleotides are bound to the filter membranes instead of the PCR products, resulting in an array of bound probes that can then be used in hybridization reactions with radioactively or nonradioactively labeled DNAs. After processing of the filters, the existence of mutations or polymorphisms in the sample can be determined by the location of the positive signals. Thus, only one filter is required to analyze a large number of different sequences. In this chapter we describe a modification *(11)* of the original procedure *(10)* for binding oligonucleotides to membrane filters, and illustrate its utility in detecting mutations in a common genetic disease, cystic fibrosis (CF) (*see* Note 1).

2. Materials

1. Oligonucleotide synthesis: In the example given in this chapter, the oligonucleotides were synthesized on a Milligen/Biosearch (Novato, CA) Model 8750 DNA synthesizer using reagents and protocols obtained from the manufacturer. The phosphoramidite nucleosides were obtained from American Bionetics, Hayward, CA.

2. 5' Reactive amino groups: Terminal, primary amino groups (aminolinkers) are added to the probe oligonucleotides during the final coupling step on the DNA synthesizer (*see* Note 2). The reagent, *N*-trifluoroacetyl-6-aminohexyl-2-cyanoethyl *N',N'*-diisopropylphosphoramidite, can be purchased from Milligen/Biosearch. Similar reagents can be obtained from Applied Biosystems (Foster City, CA). Store these derivatized oligonucleotides in water frozen at –20°C. Although not tested rigorously, some common buffers, such as Tris-HCl, may interfere with covalent attachment of the derivatized oligonucleotides by interference from the amino groups contained in the buffer molecules. All other oligonucleotides may be stored in TE buffers (TE = 10 mM Tris-HCl, 1 mM EDTA, pH 7.0–8.0) at 4°C or –20°C.

3. Biotinylated PCR products: For purposes of nonradioactive detection, a single biotin molecule is incorporated into the 5' end of one or both of the PCR primer pairs. The biotinylated PCR products are used in a chemiluminescent detection system as described below. Attachment of biotin molecules to oligonucleotides has been described in detail *(12)*. For an alternative method of biotinylating PCR products, *see* Note 3.

4. Membranes: The solid support for the probe oligonucleotides is Biodyne C, obtained from Pall Biosupport, Glencove, NY. This membrane has a derivatized surface containing a high density of anionic carboxyl groups.

5. Preactivation solution: 0.1M HCl is used to acidify the membranes just prior to addition of the activation reagent.

6. Membrane activation reagent: EDC (1-ethyl-3-(3-dimethylaminopropyl) carbodiimide hydrochloride) is prepared just before use as a 20% solution in water (Aldrich, Milwaukee, WI or Pierce Biochemicals, Rockford, IL). The EDC powder is stored dessicated at 4°C.

7. Oligonucleotide application buffer: 0.5*M* sodium bicarbonate, pH 8.4. This solution is used for applying the amino-modified oligonucleotides to the activated membrane.

8. Membrane quenching solution: 0.1*M* NaOH is used to destroy any remaining active groups on the membranes after reaction with the amino-modified oligonucleotides.

9. 10X PCR amplification buffer: 500 m*M* KCl, 200 mM Tris-HCl, 25 mM MgCl$_2$, pH 8.4. The PCR reactions are carried out in 1X PCR amplification buffer. Store 10X stock buffer at –20°C.

10. Deoxynucleotide triphosphates or dNTPs: These can be obtained from any reliable source. We routinely use neutralized, 100 m*M* stock solutions from Pharmacia, Piscataway, NJ. Store frozen at –20°C.

11. *Taq* polymerase: Thermostable DNA polymerase obtained from Perkin-Elmer/Cetus (Norwalk, CT). Store at –20°C.

12. CF PCR primer pairs: In the example given in this chapter, PCR primer pairs were synthesized to amplify regions containing mutations in the coding regions and splice junctions of exons 4, 10, 11, 20, and 21 *(13–15)*. These mutations and corresponding PCR primers are given below (*see* Fig. 1 and Note 1).

   a. A G-to-T splice mutation at the donor site at the end of exon 4, 621 + 1 G->T.

      5' TCACATATGGTATGACCCTC 3'
      5' TTGTACCAGCTCACTACCTA 3'

   b. A 3-bp deletion at codon 508 in exon 10, delta F508 .

      5' GCAGAGTACCTGAAACAGGA 3'
      5' CATTCACAGTAGCTTACCCA 3'

   c. A G-to-T change at codon 542 in exon 11, resulting in a change from Gly to a stop codon, G542X.

      5' CAACTGTGGTTAAAGCAATAGTGT 3'
      5' GCACAGATTCTGAGTAACCATAAT 3'

   d. A G-to-A change at codon 551 in exon 11 yielding a Gly to Asp change, G551D. Same primers as in step c above.

   e. A C-to-T change at codon 553 causing an Arg to stop codon change in exon 11, R553X. Same primers as in step c above.

   f. A G-to-A change at codon 1282 changing a Trp codon to a stop in exon 20, W1282X.

      5' TGGGCCTCTTGGGAAGAACT 3'
      5' CTCACCTGTGGTATCACTCC 3'

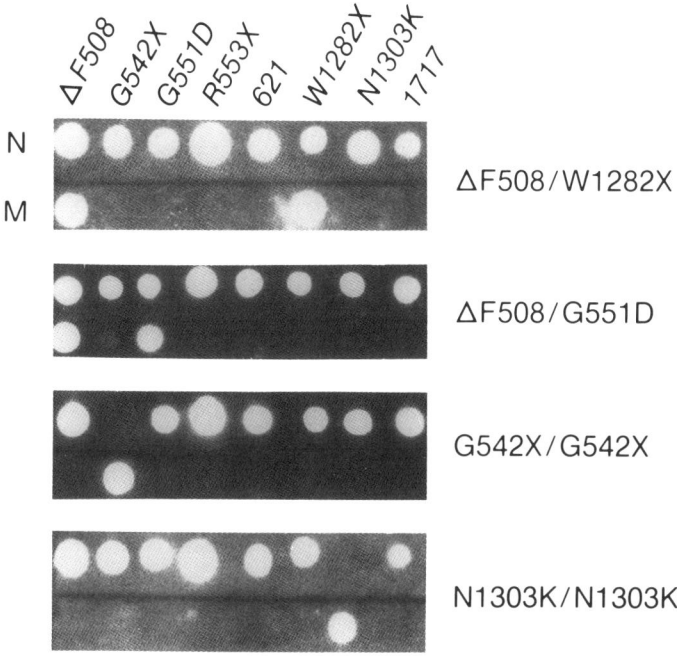

Fig. 1. Chemiluminescent detection of cystic fibrosis mutants. Four reverse dot blot strips were made with normal (N) and mutant (M) probes covalently attached in two rows as shown. The wild type and mutant probes were attached in the order shown at the top of the figure. Their sequences are given in Section 2. Each strip was hybridized with the DNAs indicated at the right of the figure. Further details are in the text. Reproduced with permission from ref. *15*.

    g. A C-to-G mutation at codon 1303 yielding an Asn to Lys change in exon 21, N1303K.

5' AAAGTATTTATTTTTTCTGG 3'

5' CTCATCTGCAACTTTCCATA 3'

13. CF oligonucleotide probes: Below are given the wild type and mutant oligonucleotide probes used as examples in the reverse dot blot experiments in this chapter. The "NH$_2$" at the 5' ends of the probes signifies the amino linkers used to attach the probes to the activated membranes. The order of the probes are as shown in Fig. 1.

Mutation	Wild Type Probe	Mutant Probe
delta F508	NH$_2$-GAAACACCAAAGATGATA	NH$_2$-GGAAACACCAATGATATT
G542X	NH$_2$-TAGTTCTTGGAGAAGGT	NH$_2$-CCTTCTCAAAGAACTA

G551D	NH$_2$-TCGTTGACCTCCACT	NH$_2$-AGTGGAGATCAACGA
R553X	NH$_2$-GAGGTCAACGAGCAAG	NH$_2$-TCTTGCTCATTGACCTC
621 + 1 G->T	NH$_2$-TTATAAGAAGGTAATACTTCC	NH$_2$-GGAAGTATTAACTTCTTATAA
W128X	NH$_2$-GCTTTCCTCCACTGTTG	NH$_2$-CAACAGTGAAGGAAAGC
N1303K	NH$_2$-AGAAAAAACTTGGATCC	NH$_2$-GGGATCCAACTTTTTTCT
1717G->A	NH$_2$-TGGTAATAGGACATCTC	NH$_2$-TGGAGATGTCTTATTAC

14. 20X SSC: 3.0M NaCl, 0.3M sodium citrate, pH 7.0.
15. 2X SSC.
16. Hybridization buffer: Filters are hybridized in 2X SSC, 0.1% SDS (sodium dodecyl sulfate).
17. Wash buffer: Filters are washed in 0.5X SSC, 0.1% SDS.
18. Chemiluminescent detection enzyme: Avidin-alkaline phosphatase is obtained from Boehringer-Mannheim, Indianapolis, IN.
19. Equilibration buffer: Buffer for use with the chemiluminescent substrate: 0.1M diethylamine, pH 10, 1 mM MgCl$_2$, and 0.02% sodium azide.
20. Chemiluminescent substrate: Admantyl 1,2-dioxetane phosphate (AMPPD) from Tropix, Bedford, MA. The stock solution is at a concentration of 10 mg/mL.
21. Chemiluminescence is recorded with Polaroid film type 84, or any suitable film such as Kodak XRP or Hyperfilm-ECL (Amersham, Arlington Heights, IL).
22. Membrane washing and stripping solution: TE + 0.5% SDS.

3. Methods

3.1. DNA Isolation and Amplification

1. Purify the PCR template DNA by any standard purification method (e.g., *see* Chapters 2 and 3 of this volume and Note 4).
2. The amplification reaction is carried out in 1X PCR buffer containing 100 µM dNTPs, 30 pmol of biotinylated primers, 10–100 ng of human template DNA, and 2 U of *Taq* polymerase. The PCR profile consists of an initial denaturation at 95°C for 5 min. The samples are then subjected to PCR with 30-s incubations at 94, 55, and 72°C for 35 cycles. *See* Note 3 for an alternative method of biotinylating PCR products. A lengthy discussion of optimization of PCR reactions is beyond the scope of this chapter, but further information can be found in Note 5 and references given therein.

3.2. Covalent Attachment of Probes to Membrane Filters

The attachment of amino modified oligonucleotides to Biodyne C membranes is a slight modification of the published protocol *(11)* (*see* Notes 2 and 6).

1. Briefly rinse the membrane strips with 0.1M HCl, blot, then soak in freshly prepared 20% EDC for 15 min.
2. Rinse the filters in deionized water and blot to remove excess water. The carboxyl groups on the membranes are activated at this point and the filters must be processed immediately (within .5 h).
3. Place the filters in a dot blot apparatus (Bio-Dot, Bio-Rad, Richmond, CA) to expedite spotting of oligonucleotides. If a dot blot apparatus is not used, place filters on a nonporous surface, such as Parafilm or plastic wrap, so that the oligonucleotide solutions are not drawn through into an absorbent surface. With the dot blot instrument, apply 2 pmol of the modified probes in 10–20 µL of bicarbonate buffer (*see* Note 6). If the apparatus is not used, a smaller volume is desirable; 2 pmol in 1–2 µL of buffer.
4. Incubate the filters for 15 min at room temperature.
5. To quench the covalent attachment reaction, rinse the membranes thoroughly for 10 min with 0.1M NaOH.
6. Wash the membranes with four to five rinses in deionized water.
7. The filters may be used immediately or air-dried for subsequent use. When dry, the membranes appear to be stable for indefinite periods.

3.3. Hybridization of PCR Products to Filters

1. Amplification reactions are commonly done in 50–100 µL volumes. Dilute 20 µL of these reactions into 0.5 mL of 2X SSC and denature the DNA by heating at 95°C for 5 min.
2. Add the denatured DNAs to 5 mL of hybridization buffer and incubate individually with the filters in sealed plastic bags at 45°C for 30 min. The hybridizations are usually done in shaking water baths for better heat transfer and circulation.
3. After hybridization, process the filters in 100 mL of wash buffer at 40°C for 10–20 min. One or two washes are usually sufficient to remove background hybridization signals. *See* Note 7 for further discussion of stringency of hybridization and washing conditions.

3.4. Chemiluminescent Detection of Hybrids

1. After hybridization and washing, incubate the filters for 30 min at room temperature in 20 mL of hybridization buffer containing 5 U of avidin-alkaline phosphatase.
2. Remove the excess avidin conjugate by two 5-min washes with the same buffer without the enzyme.
3. The filters are then soaked in 1–2 mL of the diethylamine equilibration buffer for 5 min.
4. Remove the buffer and replace with 2 mL of the same buffer containing 50 µL of a 10 mg/mL solution of AMPPD.

5. Blot the filters briefly, cover with plastic wrap, and expose immediately to Polaroid or X-ray film for 30–90 s. The exposure times will be variable, but under optimal conditions, the films are often overexposed after less than 30 s (*see* Fig. 1 for an example of chemiluminescent detection).

3.5. Reuse of Filters

The activated carboxyl group on the membranes reacts with the amino group on the oligonucleotides to form a very stable amide bond. This stability allows the filters to be reused several times, the exact number of which we have not determined.

1. After recording the hybridization results on film, wash the filters with large volumes of water or TE + 0.5% SDS.
2. The filters are then be stripped of hybrid by heating in TE + 0.5% SDS for 5–10 min. The temperature should be >20°C higher than the calculated T_m (*see* Note 7 for calculation of probe T_m). Boiling does not seem to damage the filters and stripping of the filter can even be done in a microwave oven *(11)*. Usually only one heating step is required if large volumes of buffer are used.
3. After removal of the hybrids, the filters may be reused immediately or dried for storage.

4. Notes

1. An example of the utility of the reverse dot blot method with covalently attached probes is illustrated in Fig. 1 *(15)*. Figure 1 represents an experiment where normal (N) and mutant (M) CF oligonucleotides were attached to four membrane strips and hybridized to amplified DNAs that contained mutant CF sequences. The DNA samples were obtained from patients who were heterozygous for mutations delta F508/W1282X and delta F508/G551D. A sample homozygous for G542X is also illustrated. The N1303K/N1303K sample is an artificial construct made by oligonucleotide synthesis, and mixed with normal amplified DNAs from exons 4, 10, 11, and 20. No mutant DNAs for R553X, 621, and 1717 were used in this particular experiment but were shown to hybridize to the probes in separate studies *(15)*. Their corresponding probes, however, were bound to the filters to demonstrate that the reverse dot blot test can be highly specific; i.e., there is no crosshybridization of these probes with amplified DNAs containing other mutations. Thus, with the use of a 96-space dot blot apparatus, it is possible to analyze all the known CF mutations on one filter using the reverse dot blot approach.

2. Although the sensitivity of the reverse dot blot format is sufficient for most purposes, it is sometimes desirable to have an increased detection capability. This is true in the analysis of point mutations in the RAS oncogene system. In the case of genetic diseases or HLA polymorphisms, at least 50% of the PCR product is the target sequence if a heterozygous individual is being tested. However, with RAS, the amount of the PCR product containing the mutation will depend on the proportion of tumor cells in the original tumor biopsy, and this will vary widely depending on the type of cancer and the care in which the sample is taken. We have found that addition of a "spacer arm" between the amino-linker and the oligonucleotide increases the sensitivity of the assay about four-fold (*see* Fig. 2 in ref. *11*). The spacer arm is a hexaethylene glycol-based phosphoramidite *(11,16)* that can be added to the oligonucleotide just prior to addition of the amino-linker. The increased sensitivity with the spacer arm may be owing to moving of the oligonucleotide further away from the membrane attachment site, and this may lessen steric hindrance with a concomitant increase in hybridization efficiency.

3. Biotinylation of PCR products can be accomplished by two methods; attachment of biotin to the 5' ends of the PCR primers as described in this chapter *(12)*, or by direct incorporation of the nucleotide precursor, bio-11-dUTP, into the PCR product during amplification *(15,17)*. If direct incorporation of biotin into the amplified DNA is desired, use the nucleotide analog such that its final concentration is 10- to 20-fold less than the normal precursor, TTP. In this way, a bio-dUTP will be added every tenth to twentieth base where normally a TTP would be incorporated. For example, a 200-bp PCR product with 50 thymidine bases in its sequence, will contain an average of 5 to 2.5 biotins, and this amount is more than sufficient for detection purposes by the chemiluminescent method. The biotin nucleotide analog may be obtained from Enzo Diagnostics (New York, NY), Bethesda Research Laboratories (Gaithersburg, MD) or other reliable sources.

4. In many cases absolute purity of DNA (or RNA) is not necessary for use as templates in PCR reactions. However, if "impure" templates are used, care must be taken to remove inhibitors of *Taq* polymerase in the nucleic acid preparation. For further discussion of inhibitory contaminants and easier protocols designed for PCR work, *see* sample preparation methods *(15,18,19)* and Chapters 2 and 3 of this volume.

5. The protocol given for the amplification reaction described in Section 3. is one of many possible variations on the theme, but it usually suffices for most applications where the template copy number is high (>10,000) and the target is relatively small (<500 bp). Different denaturation, an-

nealing, and extension times and temperatures can be used, but these must be tested for efficacy by the investigator. A simple modification of the protocol known as "hot start" can increase the sensitivity and specificity of the PCR reaction *(20–22)*. The increase is achieved by holding back the addition of a critical component of the reaction, such as *Taq* polymerase, until after the temperature of the PCR solution has reached a level higher than the melting temperature of the primers. This extra step inhibits random priming and primer dimerization, because DNA synthesis is blocked until stringent temperatures are reached. For further discussion of optimization of PCR reactions see recent reviews *(3,4)*, methodology handbooks *(23,24)*, and vol. 15 in this series.

6. The oligonucleotide binding capacity is very high, with more than 50 pmol required to saturate an area about the size of most 96-well dot blot manifolds. At the same time, the sensitivity is such that less than 0.1 pmol of probe per spot is sufficient to give a strong hybridization signal when the target sequence in the PCR product is in high abundance *(11)*. The combination of high capacity and sensitivity makes it possible to bind several different oligonucleotides in one spot without sacrificing detection capabilities. This may prove useful in clinical settings where there are large numbers of samples and probes to deal with. If detection only requires a "yes/no" type answer, pooling of several probes in one spot will simplify diagnostic tests.

7. The cystic fibrosis mutations can be considered somewhat of an idealized system to demonstrate the covalent attachment approach. This is because almost all the probes have an entirely different sequence from each other, resulting in very little background noise or crosshybridization. This is not the case with the RAS or HLA loci, where many of the oligonucleotides will differ in sequence by only one base *(7,8)*. Here, much more careful attention must be paid in designing the probes, and to considering the stringencies of hybridization and washes. In general, we try to use oligonucleotides that are long enough to hybridize uniquely to DNA, yet short enough to allow destabilization of a hybrid containing a 1-bp mismatch *(9)*. Usually, probes with a length of 15–20 bases fill this requirement, but in some cases the "trial and error" method is necessary to obtain a usable sequence. Try to place the mismatch in the center of the probe for maximum destabilization, and avoid G:T mismatches if possible, since they seem to have only a small effect on duplex stability *(9)*. If a G:T mismatch is unavoidable on one strand, simply switch the probe to the opposite strand, and this will give a stronger C:A mismatch. All probes should have the same T_m, or at least within 2°C of each other. The formula 4°C for each G or C, and 2°C for an A or T will give you a rough estimate of the T_m of the probes *(25)*. Thus,

a 16-mer with eight Gs or Cs and eight As or Ts will have a T_m of 48°C. Again, the values will be approximate and optimization may require adding or subtracting a base or two to equalize hybridization conditions for a particular set of probes.

References

1. Saiki, R. K., Scharf, S., Faloona, F., Mullis, K. B., Horn, G. T., Erlich, H. A., and Arnheim, N. (1985) Enzymatic amplification of beta-globin genomic sequences and restriction site analysis for diagnosis of sickle cell anemia. *Science* **230,** 1350–1354.
2. Saiki, R. K., Gelfand, D. H., Stoffel, S., Scharf, S. J., Higuchi, R., Horn, G. T., Mullis, K. B., and Erlich, H. A. (1988) Primer-directed enzymatic amplification of DNA with a thermostable DNA polymerase. *Science* **239,** 487–491.
3. Bloch, W. (1991) A biochemical perspective of the polymerase chain reaction. *Biochemistry* **30,** 2735–2747.
4. Erlich, H. A., Gelfand, D., and Sninsky, J. J. (1991) Recent advances in the polymerase chain reaction. *Science* **252,** 1643–1651.
5. Bos, J. L. (1989) RAS oncogenes in human cancer: a review. *Cancer Res.* **49,** 4682–4689.
6. Roberts, L. (1990) Cystic fibrosis diagnosis. *Science* **250,** 1076,1077.
7. Erlich, H., Buguwan, D., Begovich, A. B., Scharf, S., Griffith, R., Saiki, R., Higuchi, R., and Walsh, P. S. (1991) HLA-DR, DQ and DP typing using PCR amplification and immobilized probes. *Eur. J. Immunogenetics* **18,** 33–55.
8. Verlaan-de Vries, M., Bogaard, M. E., van den Elst, H, van Boom, J. H. van der Eb, A. J., and Bos, J. L. (1986) A dot-blot screening procedure for mutated RAS oncogenes using synthetic oligonucleotides. *Gene* **50,** 313–320.
9. Ikuta, S., Takagi, Wallace, R. B., and Itakura, K. (1987) Dissociation kinetics of 19 base paired oligonucleotide -DNA duplexes containing different single mismatched base pairs. *Nucleic Acid Res.* **15,** 797–811.
10. Saiki, R. K., Walsh, P. S., Levenson, C. H., and Erlich, H. A. (1989) Genetic analysis of amplified DNA with immobilized sequence-specific oligonucleotide probes. *Proc. Natl. Acad. Sci. USA* **86,** 6230–6234.
11. Zhang, Y., Coyne, M. Y., Will, S. G., Levenson, C. H., and Kawasaki, E. S. (1991) Single-base mutational analysis of cancer and genetic diseases using membrane bound modified oligonucleotides. *Nucleic Acid Res.* **19,** 3929–3933.
12. Levenson, C. and Chang, C.-A. (1990) Nonisotopically labeled probes and primers, in *PCR Protocols* (Innis, M. A., Gelfand, D. H., Sninsky, J. J., and White, T. J., eds.), Academic, San Diego, CA, pp. 99–112.
13. Zielenski, J., Rozmahel, R., Bozon, D., Kerem, B. S., Grelczack, Z., Riordan, J. R., Rommens, J., and Tsui, L. C. (1991) Genomic DNA sequence of the cystic fibrosis transmembrane conductance regulator (CFTR) gene. *Genomics* **10,** 214–228.
14. Osborne, L., Knight, R. Santis, G., and Hodson, M. (1991) A mutation in the second nucleotide binding fold of the cystic fibrosis gene. *Am. J. Hum. Genet.* **48,** 608–612.

15. Chehab, F. F. and Wall, F. (1992) Detection of multiple cystic fibrosis mutations by reverse dot blot hybridization: a technology for carrier screening. *Hum. Genetics* **89,** 163–168.

16. Levenson, C. H. and Chang, .C-A. (1990) U.S. Patent # 4,914,210.

17. Lo, Y.-M. D., Mehal, W. Z., and Fleming, K. A. (1990) Incorporation of biotinylated dUTP, in *PCR Protocols* (Innis, M. A., Gelfand, D. H., Sninsky, J. J., and White, T. J., eds.), Academic, San Diego, CA, pp. 113–118.

18. Higuchi, R. (1989) Simple and rapid preparation of samples for PCR, in *PCR Technology* (Erlich, H. A., ed.), Stockton, New York, pp. 31–38.

19. Kawasaki, E. S. (1990) Sample preparation from blood, cells, and other fluids, in *PCR Protocols* (Innis, M. A., Gelfand, D. H., Sninsky, J. J., and White, T. J., eds.), Academic, San Diego, CA, pp. 146–152.

20. Nuovo, G. J., Gallery, F., MacConnell, P., Becker, J., and Bloch, W. (1991) An improved technique for the *in situ* detection of DNA after polymerase chain reaction amplification. *Am J. Pathol.* **139,** 1239–1244.

21. D'Aquila, R. T., Bechtel, L. J., Videler, J. A., Eron, J. J., Gorczyca, P., and Kaplan, J. C. (1991) Maximizing sensitivity and specificity of PCR by pre-amplification heating. *Nucleic Acids Res.* **19,** 3749.

22. Chou, Q., Birch, D., Russell, M., Raymond, J., and Bloch, W. (1992) Prevention of pre-PCR mispriming and primer dimerization improves low copy number amplification. *Nucleic Acids Res.* **20,** 1717–1723.

23. Erlich, H. A. (1989) *PCR Technology: Principles and Applications for DNA Amplification.* Stockton, New York.

24. Innis, M. A., Gelfand, D. H., Sninsky, J. J., and White, T. J. (eds.) (1990) *PCR Protocols: A Guide to Methods and Applications.* Academic, San Diego, CA.

25. Miyada, C. G. and Wallace, R. (1987) Oligonucleotide hybridization techniques. *Methods Enzymol.* **154,** 94–107.

CHAPTER 34

RAPD Assay

A Novel Technique for Genetic Diagnostics

Joseph P. del Tufo and Scott V. Tingey

1. Introduction

The analysis of DNA-based polymorphisms is integral to the construction of molecular genetic maps. Currently, the method of choice is the restriction fragment length polymorphism (RFLP) assay. The RFLP method only allows the detection of DNA sequence polymorphisms by Southern blot hybridization and is, in general, a time-consuming and labor-intensive method. Recently, a new DNA polymorphism assay termed RAPD for random amplified polymorphic DNA *(1,2)*, has been developed that allows the detection of DNA sequence polymorphisms using single short primers of an arbitrary nucleotide sequence in a DNA amplification assay. Normally, primers of ten nucleotides in length are used. Since polymorphisms are simply detected as DNA segments that amplify from one parent but not the other, and they are inherited through classical Mendelian genetics, they can be used to construct genetic maps in an assortment of species.

RAPD markers provide an opportunity to increase the efficiency of genetic map creation. Reiter et al. *(3)* demonstrated this utility by using RAPD markers to create a complete genetic map of *Arabidopsis thaliana*, containing 250 RAPD markers, in only 4 mo.

From: *Methods in Molecular Biology, Vol. 28:*
Protocols for Nucleic Acid Analysis by Nonradioactive Probes
Edited by: P. G. Isaac Copyright ©1994 Humana Press Inc., Totowa, NJ

Since polymorphisms are noted by the presence or absence of amplification products from a single allele, the RAPD technique tends to provide only dominant markers. Individuals containing two copies of one allele are not distinguished by amplification from those with only one copy. Dominant markers provide little linkage information for markers linked in repulsion. Therefore, when mapping, it is advisable to only work with markers linked in coupling, e.g., in a backcross or recombinant inbred population, haploid, or gametophytic tissue or, alternately, in an F2 population with markers amplified from only one parent.

RAPD markers are especially practical when used in techniques that are designed to reveal polymorphism in targeted regions of a genome. Near isogenic lines *(4,5)* can be screened to identify regions of the genome introgressed from the donor parent. Bulked segregant analysis (BSA) *(6)* has been developed as an assay for quickly identifying markers linked to any specific region of any genome. In theory, two bulked DNA samples are gathered from a segregating population, where each bulk is composed of individuals identical for a specific trait or region, but randomized in all unlinked regions. Since many segregating individuals are used to generate the bulks, there is only a minimal chance that regions unlinked to the target region will differ between the bulked samples. A similiar approach uses genetic mapping information to create bulks based on genotype. In this way Giovannoni et al. *(7)* and Reiter et al. *(3)* were able to target RAPD markers to a defined region of the tomato and *Arabidopsis* genome, repectively. Combined with the RAPD assay, screening near isogenic lines or using BSA can quickly add markers to a specific region of the genome.

Using an arbitrary primer as short as 5 nucleotides Gresshoff et al. *(8)* were able to produce a detailed and relatively complex DNA profile. This approach, termed DNA Amplification Fingerprinting (DAF), which utilizes silver staining of DNA and acrylamide gels, promises to obtain fingerprints with greater ease than is possible by conventional techniques.

The RAPD protocol is relatively quick and easy, fluorescence is used in lieu of radioactivity, and the assay saves time, effort, and the myriad risks and precautions associated with Southern blot technology.

2. Materials

1. Approximately 20 ng of genomic DNA per assay (*see* Note 1).
2. A single 10-base oligodeoxynucleotide primer of random sequence, containing 50–70% G + C. 20 pg of primer are required for each assay (*see* Note 1).
3. 10X *Taq* buffer: 100 m*M* Tris-HCl, pH 8.3, 500 m*M* KCl, 19 m*M* MgCl$_2$, 0.1% gelatin.
4. 20 m*M* MgCl$_2$.
5. 2 m*M* Solution of all four deoxynucleotide triphosphates, dNTPs (dATP, dCTP, dGTP, dTTP).
6. *Taq* Polymerase: 5 U/µL (Perkin-Elmer/Cetus, Norwalk, CT)
7. Reaction premix: Make up enough mixture for all of the samples to be assayed. For a single 25-µL reaction mix 15.8 µL of sterile distilled water, 2.5 µL of 10X *Taq* buffer, 1.25 µL of 2 m*M* dNTPs, 0.25 µL of 20 m*M* MgCl$_2$ (*see* Note 2) and 0.2 µL of 5 U/µL *Taq* polymerase. Mix thoroughly.
8. Mineral oil.

3. Method

1. Pipet 20 µL of reaction premix into each tube, add approx 20 ng of template DNA, and 20 pg of primer (*see* Notes 1, 3, and 4) to bring the reaction volume to 25 µL. (The final reaction conditions are 10 m*M* Tris-HCl, pH 8.3, 50 m*M* KCl, 2.1 m*M* MgCl$_2$ [*see* Note 2], 0.01% gelatin, 100 µ*M* each dNTP, 0.2 µ*M* primer, and 1 U of *Taq* polymerase/ 25 µL reaction.)
2. Overlay each reaction with 50 µL mineral oil.
3. This reaction is placed in a Perkin-Elmer/Cetus thermocycler or equivalent for 45 cycles. Each cycle is composed of 1 min at 94°C (denaturation step), 1 min at 35°C (annealing step), and 2 min at 72°C extension step). The final cycle is culminated by 5 min at 72°C and then held at 4°C until assayed (*see* Note 5).
4. The sizes of the amplification products are determined by gel electrophoresis in a 1.4% agarose gel and visualized by staining with ethidium bromide (*see* Fig. 1 and Note 6).

4. Notes

1. The concentrations of DNA and primers are such that the combined volume of 20 ng of DNA and 20 pg of primer is 5 µL.
2. Fluctuation in the final concentration of MgCl$_2$ above or below 2.1 m*M* can often cause a marked decrease in amplification products. The con-

A B

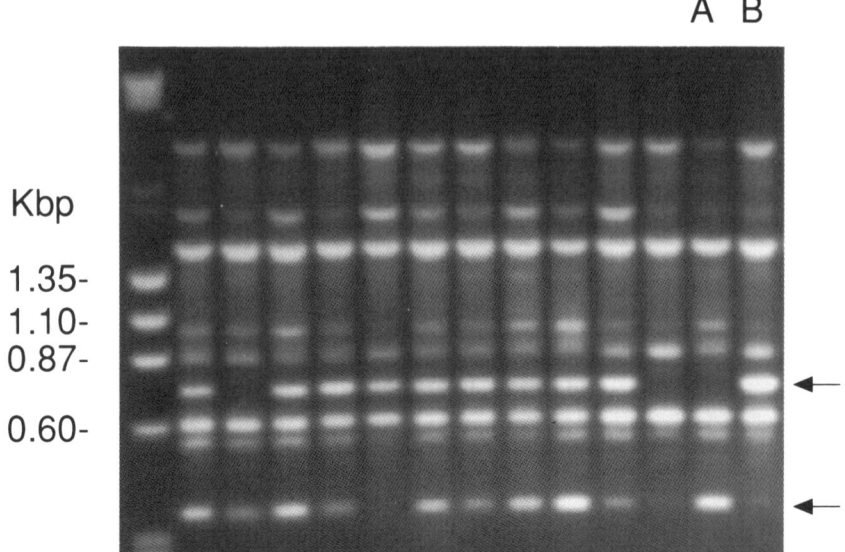

Fig. 1. DNA was isolated from 11 F2 individuals segregating from a cross of two *Glycine max* (soybean) cultivars (A and B). Each DNA sample was amplified with a single 10-mer oligodeoxynucleotide primer (5'-AGCGTGTCTG). The amplification products were separated by electrophoresis in a 1.4% agarose gel and visualized by staining with ethidium bromide. The arrows at the right of the gel point to polymorphisms between parents A and B that are segregating in the F2 generation. The mol-wt standards at the left of the gel are from a *Hae* III digest of ØX 174 DNA.

centration of MgCl₂ that will produce the most consistent results will usually depend on the quality of the template DNA.

3. The ratio of primer to template DNA in the RAPD reaction is critical. A shift in concentration of either the primer or template DNA can result in the inconsistent amplification of discrete loci, increased background, or failure of amplification. Template DNA concentration should be carefully titered against a standard concentration of primer. This will reveal the concentration of DNA that gives the most reproducible amplified products. A titration should be performed when using DNA from different organisms, or when using DNA of varying quality, including DNA isolated with different extraction procedures. It is not necessary to retiter each DNA sample as long as the concentration of DNA and the quality of DNA remains constant.

4. Primer quality is important. If the amplification reaction is initiated with a primer containing significant levels of premature termination products, the RAPD reaction will usually not work. If care is taken during the synthesis and storage of the oligodeoxynucleotide, purification is usually not required beyond deprotection and desalting of the primer.
5. In order to obtain reproducible results, one should maintain consistent thermocycling conditions. Fluctuation of cycle time or number of cycles, or the use of different brands of thermocyler will often result in different amplification products.
6. Occasionally one notices a hazy smear partially or completely obscuring the amplified bands on the agarose gel. This is usually caused by a failure to saturate the DNA template with primer, and can be corrected by adjusting the ratio of primer to template DNA.

References

1. Williams, J. G. K., Kubelik, A. R., Livak, K. J., Rafalski, J. A., and Tingey, S. V. (1990) DNA polymorphisms amplified by arbitrary primers are useful as genetic markers. *Nucleic Acids Res.* **18**, 6531–6535.
2. Welsh, J. and McClelland, M. (1990) Fingerprinting genomes using PCR with arbitrary primers. *Nucleic Acids Res.* **18**, 7213–7218.
3. Reiter, R. S., Feldman, K. A., Williams, J. G. K., Rafalski, J. A., Tingey, S. V., and Scolnik, P. A. (1992) Genetic linkage of the arabidopsis genome: methods for mapping with recombinant inbreds and random amplified polymorphic DNAs. *Proc. Natl. Acad. Sci. USA* **89**, 1477–1481.
4. Paran, I., Kesseli, R., and Michelmore, R. W. (1991) Identification of restriction fragment length polymorphisms and random amplified polymorphic DNA markers linked to downy mildew resistance genes in lettuce, using near-isogenic lines. *Genome* **34**, 1021–1027.
5. Martin, G. B., Williams, J. G. K., and Tanksley, S. D. (1991) Rapid identification of markers linked to a pseudomonas resistance gene in tomato by using random primers and near isogenic lines. *Proc. Natl. Acad. Sci. USA* **88**, 2336–2340.
6. Michelmore, R. W., Paran, I., and Kesseli, R. V. (1991) Identification of markers linked to disease resistance by bulked segregant analysis: a rapid method to detect markers in specific genomic regions using segregating populations. *Proc. Natl. Acad. Sci. USA* **88**, 9828–9832.
7. Giovannoni, J. J., Wing, R. A., Ganal, M. W., and Tanksley, S. D. (1991) Isolation of molecular markers from specific chromosomal intervals using DNA pools from existing mapping populations. *Nucleic Acids Res.* **19**, 6553–6558.
8. Caetano-Anolles, G., Bassam, B. J., and Gresshoff, P. M. (1991) High resolution DNA amplification fingerprinting using very short arbitrary oligonucleotide primers. *Biotechnology* **9**, 553–557.

CHAPTER 35

Nonradioactive Oligonucleotide Probes for Detecting Products of the Ligase Chain Reaction

Jon Kratochvil and Thomas G. Laffler

1. Introduction

The ligase chain reaction (LCR) is a novel DNA amplification system developed jointly by Abbott Laboratories (North Chicago, IL) and Biotechnica International, Inc. (Kansas City, MO) *(1)*. By amplification with specially haptenated detection probes, LCR can detect less than 100 target DNA molecules in a nonradioactive assay format.

LCR uses two probe pairs that employ the specific nucleic acid sequence of a target molecule to direct geometric amplification *(2)*. The first probe set contains two contiguous nucleic acid sequences, A and B, that are complementary to the target region. The second set, A' and B', are complementary to the first probe set and identical to the target region. The probes are added in excess to the sample and after a heat denaturing step are allowed to hybridize to complementary target regions at a temperature near the T_m of the complementary probes sets, A-A' or B-B'. This step determines the specificity of the LCR. Only probes that have hybridized to the continuous target sequences will abut one another to generate a ligatable substrate. These adjacent probes are then joined by a thermostable ligase (i.e., from *Thermus thermophilus*). Repeated cycles of heating and cooling melt

From: *Methods in Molecular Biology, Vol. 28:*
Protocols for Nucleic Acid Analysis by Nonradioactive Probes
Edited by: P. G. Isaac Copyright ©1994 Humana Press Inc., Totowa, NJ

the double-stranded ligation products and allow unreacted probes to reanneal to the single stranded target molecules. The ligated products act as target molecules for the complementary probe sets in subsequent cycles, resulting in exponential amplification of the original nucleic acid target sequence to readily detectable levels (*see* Fig. 1).

The nonradioactive detection assay utilizes two specific ligands, biotin and fluorescein, that are coupled to the termini of the probes distal from the ligation junction. Probe set, A and A', carry the capture molecule fluorescein at their 5' and 3' distal ends, respectively. Probe set, B and B', have the signal-generating group, biotin, at their corresponding ends. Following amplification, the LCR ligation products will carry a different binding ligand at each end. These bidentate molecules enable detection by sandwich immunoassays in a number of formats (*see* Note 1). We commonly use an automated microparticle capture enzyme immunoassay (MEIA). Probes containing the fluorescein moiety are captured by microparticles coated with anti-fluorescein antibody. Unbound, unligated biotinylated probe is removed by a washing step. An antibiotin antibody conjugated to alkaline phosphatase is then introduced. This conjugate will bind to the biotinylated LCR-product molecules captured by the solid phase. Following a wash, a fluorescent detection signal is generated by reaction of the retained alkaline phosphatase, with the fluorogenic substrate, methyl umbelliferyl phosphate. This signal is read in the Abbott IMx®, an enzyme immunoassay analyzer that automates the entire microparticle-based detection assay *(3)*. Only molecules linking both the fluorescein and biotin derivatives (i.e., the ligated products) can generate a signal. Included in this chapter is a protocol for manual haptenation of the oligonucleotide probes yielding the most consistent level of haptenation. Since many laboratories do not have access to an IMx analyzer, we describe a simple, immuno-chromatographic detection system that requires no sophisticated equipment *(4)*.

2. Materials

1. 15 A_{260} U of each of four oligonucleotide probes (oligos) containing the exact sequence of or exact complementary sequence to the target molecule to be detected (*see* Notes 2–4).
2. Fluorescein-5-isothiocyanate (Molecular Probes, Inc., Eugene, OR).

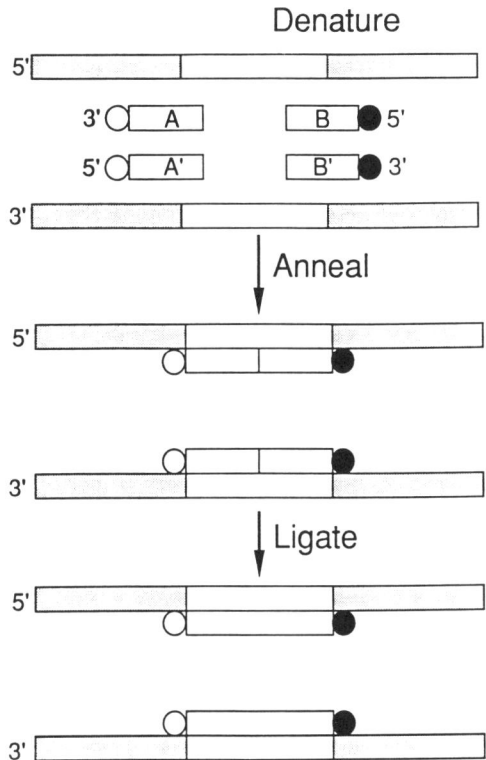

Fig. 1. Principle of LCR. Probes are added in excess to the DNA target sample and the nucleic acid probes and target are denatured at 95°C. Probes anneal to the target at a temperature near the T_m of the probes and are ligated by a thermostable ligase, resulting in a doubling of target sequences. Further cycles of denaturing at 85°C, annealing and ligation result in the exponential amplification of detectable product. Probes A and A' are coupled to fluorescein (open circles) and probes B and B' are tagged with biotin (closed circles).

3. Biotin-XX-NHS ester (Clonetech, Palo Alto, CA).
4. 0.1*M* Sodium borate, pH 9.2.
5. 0.1*M* Sodium phosphate buffer, pH 7.2.
6. *N,N* dimethylformamide (**Caution:** This is a potential carcinogen and should be dispensed in a fume-hood).
7. Four 5-mL NAP-5 columns (Pharmacia, Uppsala, Sweden).
8. 0.3*M* Sodium acetate (NaOAC), pH 5.5 with glacial acetic acid.
9. 100% Ethanol.

10. Deionized formamide (**Caution:** This is a potential carcinogen and should be dispensed in a fume-hood).
11. 5X TBE: 445 mM trizma base, 445 mM borate, 10 mM EDTA.
12. Polyacrylamide gel solution: To prepare a 300-mL acrylamide solution for oligonucleotide resolution add the following to a 500-mL graduated cylinder: 100 mL 36% acrylamide in distilled water (19:1 acrylamide:bis) (**Caution:** Acrylamide is a potent neurotoxin! Use gloves and a face mask when handling liquid or when weighing out powder), 60 mL 5X TBE, 144 g urea. Bring the volume to 300 mL with distilled water (*see* Note 5).
13. 5% Sodium dodecyl sulfate (SDS, also known as sodium lauryl sulfate) in distilled water (w/v) (**Caution:** This is an irritant. Wear gloves and face mask when weighing out the powder).
14. 1% Agarose in 1X TBE.
15. 10% Ammonium persulfate.
16. TEMED.
17. Oligonucleotide elution buffer: 100 mM Tris-HCl, pH 8.0, 500 mM NaCl, 5 mM EDTA.
18. Four 1-mL Sep-Pak C18 push columns (Waters Associates, Milford, MA).
19. 100% Methanol.
20. 50% Methanol/50 mM Triethylamine acetate (TEAA).
21. LCR reaction mix: 20 mM Trizma base, 10 mM MgCl$_2$, 0.1% Triton-X100, 25 mM KOH, 10 mM DTT, 0.6 mM NAD, adjust the pH to pH 7.6 with glacial acetic acid and add haptenated oligonucleotides. This reaction mix is normally made up in bulk so that 45 μL contains 7.5 × 10^{11} mol of probes A' and B, and 5 × 10^{11} mol of probes A and B'.
22. DNA ligase from *Thermus thermophilis* (Epicentre, Madison, WI).
23. 3% SeO$_2$ (w/v).
24. 1% Ascorbic acid (w/v).
25. Nitrocellulose: 5 μm (Schleicher and Schuell, Dassel, Germany).
26. TBS: 0.1M Tris-HCl, pH 7.8, 0.9% NaCl.
27. 2.66% Casein in 1.33X TBS.
28. Phosphate buffered saline: 30 mM NaCl, 20 mM sodium phosphate buffer, pH 7.3.
29. Bovine serum albumin, fraction V (BSA).
30. Antifluorescein polyclonal antibody (Boehringer Mannheim, Mannheim, Germany): 0.5 mg/mL dissolved in 0.1% BSA in TBS.
31. Mouse antibiotin monoclonal antibody (Sigma Chemicals, St. Louis, MO): 4.6 mg/mL.
32. 1.7-mL and 650-μL siliconized Slickseal™ microfuge tubes (National Scientific, San Rafael, CA).
33. Sterile distilled water.

3. Methods

3.1. Preparation of Haptenated Probes

Unless stated otherwise, all haptenation reactions are run at room temperature.

1. Resuspend 15 A_{260} U of each oligo in 10 µL distilled water in a sterile 1.7-mL microfuge tube.
2. Dissolve 10 mg of fluorescein-5-isothiocyanate in 1 mL of sodium borate solution in a 1.7-mL siliconized microfuge tube.
3. Dissolve 10 mg of biotin-XX-NHS ester in 500 µL of *N,N* dimethyl-formamide in a glass tube. This may need vortexing to fully dissolve the biotin (*see* Note 6).
4. Add 500 µL of the fluorescein-5-isothiocyanate solution to oligonucle-otides A and A', mix well and place them in the dark for at least 12 h.
5. Add 240 µL of 0.1*M* sodium phosphate solution to probes B and B' and mix well.
6. Add 250 µL of biotin-XX-NHS ester solution to probes B and B', mix well and place them in the dark at least 12 h.
7. The following day wash four NAP-5 columns with 10 mL distilled water.
8. Load each oligo solution on a separate NAP-5 column.
9. Elute the oligos from the column into a 1.7-mL microfuge tube using 1 mL distilled water. (The NAP-5 column contains a size exclusion media that removes unreacted ligands from oligonucleotides.)
10. Dry the oligonucleotides to near dryness in a SpeedVac, resuspend them in 200 µL 0.3*M* sodium acetate, pH 5.5, and precipitate them with 2.5 vol 100% ethanol at –70°C for 0.5 h.
11. Sediment each precipitated oligonucleotide in a microcentrifuge at 13,000*g* for 10 min at 4°C, pour off the supernatant and allow the oligonucleotide pellet to air-dry for 1 h or dry for a few minutes in a SpeedVac.
12. Resuspend the pellets in 50 µL of deionized formamide and hold the tubes in a boiling water bath for 5 min.

3.2. Preparation of Polyacrylamide Gel

1. Filter the 36% acrylamide solution through a 1000-mL Nalgene CN 0.45-µm filter apparatus.
2. Prepare the acrylamide gel apparatus sandwich using clean 30 × 40 cm glass plates and 1.6-mm spacers. Wash the plates with successive washes of 5% SDS solution, distilled water, and 100% ethanol. Dry the plates with a lint-free towel. Clamp the apparatus together with bookbinder clamps and seal the edges with 1.0% agarose in 1X TBE (*see* Notes 7 and 8).

3. Add 750 µL of 10% ammonium persulfate and 100 µL of TEMED to 300 mL of the acrylamide solution. Slowly and evenly pour this solution into the gel sandwich to avoid generating bubbles. Place a 16-well preparative comb into the top of the gel and allow the gel to polymerize for 1 h (*see* Note 9).

4. Remove the comb and prerun the gel in a 1X TBE running buffer at 1000 V (25 V/cm) for 1 h. Every 15 min, disconnect the power supply and wash the wells with 1X TBE in a 10-mL syringe fixed to a 23-g needle to remove the excess urea (*see* Note 10).

3.3. Purification of Haptenated Oligonucleotides

1. Load the oligo samples onto the prerun gel and run the gel at 1000 V until fluoresceinated bands are 5 cm from the bottom (approx 6 h) (*see* Note 11).

2. Visualize the bands by placing the gel on an intensifying screen and shadowing with a UV light source. The fluoresceinated bands can be observed by shadowing at 320 nm. The biotinylated nucleic acid bands can be visualized by shadowing at 260 nm (*see* Note 12).

3. Cut out each haptenated oligonucleotide band with a clean razor blade, place in 3 mL of elution buffer in a 15-mL conical centrifuge tube, and allow the oligo-nucleotides to diffuse from the gel slice at 60°C for at least 6 h.

4. Attach a Sep-Pak column to the base of a 10-mL luer-lock syringe and wash with 10 mL 100% methanol followed by 10 mL distilled water.

5. Load each oligonucleotide diffusion mixture onto the columns and desalt by washing with 20 mL distilled water.

6. Elute each oligonucleotide into three 1.7-mL siliconized microfuge tubes with 3 mL of a solution of 50% methanol/50 mM TEAA.

7. Dry down the probe eluant in a SpeedVac, pool the respective oligos into one tube each after dissolving the pellets in 33.3 µL of 0.3M sodium acetate, pH 5.5, and then precipitate them with 2.5 vol 100% ethanol at −70°C for 0.5 h.

8. Spin down the precipitated haptenated oligonucleotides in microcentrifuge at 13,000g for 10 min at 4°C, dry the pellets in a SpeedVac, and resuspend them in 100 µL of sterile distilled water. Allow at least 1 h for the DNA to fully rehydrate.

9. Quantify the oligonucleotide by absorbance at 260 nm. Extinction coefficients may be calculated using the Oligo 4.0 program by Rychlik (*5*) (*see* Note 13). The resulting yield is usually between 1×10^{14} mol/µL for the biotinylated probes and 5×10^{14} mol/µL for fluoresceinated probes.

3.4. LCR Amplification

1. Assemble the reaction mixtures in siliconized 500-µL tubes. Each reaction mix contains 7.5×10^{11} mol of probes A' and B, and 5×10^{11} mol of probes A and B' in 45 µL of the LCR reaction mix.

2. Add 3 µL of the DNA sample to be tested (5×10^2 to 5×10^5 mol/reaction).
3. Overlay tubes with a drop of mineral oil and heat to 100°C for 3 min, hold at 85°C for 1 min, and at 50°C for an additional 1 min. Add 2 µL of *Thermus thermophilis* DNA ligase to bring the final volume to 50 µL (*see* Note 14).
4. Place the tubes in a Coy "TempCycler" programmed for 30 cycles of 30 s at 85°C then 20 s at 50°C (*see* Note 15).
5. After the cycling is completed, remove the aqueous layer from beneath the oil.
6. Quantify the amplification products by the sandwich immunoassay protocol of your choice or use the immunochromatographic technique detailed below.

3.5. Solid Phase Immunoassay

3.5.1. Preparation of Colloidal Selenium

1. Add 2 mL of 3% SeO_2 and 4.5 mL of freshly prepared 1% ascorbic acid to 200 mL of boiling distilled water.
2. Reflux this solution for 10 min and cool to room temperature. Centrifuge the resulting reddish brown solution for 20 min at 2500g and 4°C.
3. Resuspend the pellet in 100 mL of distilled water. Read the absorbance at 540 nm in a spectrophotometer, and add distilled water to the solution so that its A_{540} is 15.

3.5.2. Conjugation of Antibiotin Monoclonal Antibody with Colloidal Selenium

Prepare this on the day of use.

1. Adsorb monoclonal antibody against biotin onto colloidal selenium by mixing 5 µL of goat antibiotin antibody (4.6 mg/mL) with 0.4 mL of phosphate buffered saline and then adding 5 mL of the above selenium colloid.
2. Stir 10 min at 26°C.
3. Add 0.108 g of powdered BSA to a final concentration of 2% (w/v).
4. Add 16.2 mL of 2.66% casein, 1.33X TBS. The concentration of the working conjugate is 2% casein, 1X TBS, with an A_{540} of 3.75.

3.5.3. Preparation of Solid Phase Medium

Spot 4.0 mm × 4.0 cm nitrocellulose strips (*see* Note 16) with 0.2 µL of 0.5 mg/mL antifluorescein dissolved in 0.1% BSA in TBS 1 cm from the bottom of the strip. Allow the spots to air-dry at room temperature. This zone will capture all the fluoresceinated oligonucleotides. The strips may be stored dessicated at room temperature indefinitely.

3.5.4. Resolution of LCR Products

1. Mix the 5 µL of the LCR product with 40 µL of the selenium conjugate. Incubate for 5 min at room temperature.
2. Place the bottom of the nitrocellulose strips in this incubation solution and allow the selenium colloid-antibiotin oligonucleotide complex to migrate up the strip (approx 5 min). Only LCR products that contain both fluorescein and biotin will be captured at the antifluorescein zone and will be indicated by a red color.

4. Notes

1. Some other assay formats include using antifluorescein coated microtiter wells, polystyrene beads, or antifluorescein spotted nitrocellulose strips *(6)*.
2. We have found that the optimal length of the probes range from 15 to 30 bases each and that all probes in an assay should preferably be of approximately equal length. It is also preferred that the corresponding probe pairs have similar T_m.
3. Probes A and B' must have amino-modifiers at their 5' termini and B and A' must have these modifiers at their 3' termini and phosphates at their 5' termini. These modified oligonucleotides can be ordered from a variety of nucleic acid synthesis laboratories.
4. Oligonucleotides can also be ordered that have already been derivatized at the appropriate ends with biotin or fluorescein using control pore glass (CPG) coupled with a ligand for the 3' termini and a phosphoramidite coupled with a ligand for the 5' termini. However, we have found the yields of these oligonucleotides to have been significantly lower than our manually haptenated lots. More importantly, fluorescein CPG does not react well with its corresponding oligonucleotide and gives poor, unreliable coupling.
5. This reaction is highly endothermic and the concentration of urea will be near the saturation point. It is helpful to maintain a temperature of 30°C to increase the solubility of the urea.
6. A water soluble biotin ligand is now available from several vendors that may alleviate the solubility and handling problems of the currently described biotin ligand. However, its efficacy in this protocol is yet to be determined.
7. Bubbles will form in the gel during pouring if the plates are not cleaned scrupulously.
8. Many companies provide apparati that do not need to be clamped and sealed with agarose.

9. If the wells are malformed, the gel may have polymerized too quickly, forcing water to the top of the gel. If this is the case it will be necessary to reduce the ammonium persulfate and TEMED concentrations.

10. This is a crucial step. Excess urea in the wells will impede the migration of samples into the gel. The oligonucleotides will run much more discretely and uniformly when they are loaded into clean wells.

11. Fluoresceinated oligo bands should be visible after entering the gel. If only biotinylated probes are run on a gel, visible markers must be loaded into a separate well to track the migration of the probes.

12. Fluorescein ligands will add approx 1.5 bases to the oligonucleotide migration in the gel and biotin will add 3 bases. This easily allows the differentiation between haptenated and unhaptenated oligonucleotides.

13. We have found that LCR efficiency is highly dependent on the amount of probe per reaction. Typically, the optimal probe concentrations will range between 5×10^{11} to 1×10^{12} mol of each probe per sample reaction. Owing to this narrow working window for the probe sets, it is necessary to quantify the oligonucleotide concentration as accurately as possible. The Oligo 4.0 program is available commercially from National Biosciences Inc., Plymouth, MN.

14. The LCR conditions shown serve as one, specific example. For individual applications, conditions must be optimized. We have found the critical optimization factors to be probe concentration, annealing temperature, cycle number, and ligase concentration. Any commercially available thermophilic DNA ligase should be suitable with a compatible buffer. We have had good success with ligase purified from *Thermus thermophilis*. Since there is no universally accepted enzyme unit definition, the optimal amount of enzyme to use must be determined experimentally.

15. Any commercially available thermal cycler should suffice. The optimal thermal cycling program will need to be determined for each type of cycler.

16. The production of nitrocellulose can be complicated. Although any manufacturer of nitrocellulose may be adequate, Schleicher and Schuell consistently manufacture a high quality product.

References

1. Backman, K. C. and Wang, C. J. (1989) Method for Detecting a Target Nucleic Acid Sequence. European Patent Application 320308.

2. Bond, S., Carrino, J., Hampl, H., Hanley, K., Rinehardt, L., and Laffler, T. (1990) New methods for detecting HPV: papillomaviruses, in *Human Pathology Recent Progress in Epidermoid Precancers*, vol. 78 (Monsenego, J., ed.), Serono Symposia Publications.

3. Fiore, M., Mitchell, J., Doan, T., Nelson, R., Winter, G., Grandone, C., Zeng, K., Haraden, R., Smith, J., Harris, K., Leszczynski, J., Berry, D., Safford, S., Barnes, G., Scholnick, A., and Ludington, K. (1988) The Abbott IMx™ automated benchtop immunochemistry analyzer system. *Clin. Chem.* **34,** 1726–1732.
4. Yost, D., Russell, J., and Yang, H. (1990) Non-Metal Colloidal Particle Immunoassay: United States Patent 4,954,452.
5. Rychlik, W. and Rhoads, R. E. (1989) A computer program for choosing optimal oligonucleotides for filter hybridization sequencing and in-vitro amplification of DNA. *Nucleic Acids Res.* **17,** 8543–8552.
6. Laffler, T. and Bouma, S. (1989) Detection and Amplification of Target Nucleic Acid Sequences. European Patent Application 0357011A2.

CHAPTER 36

Nucleic Acid Sequence-Based Amplification (NASBA™)

Larry Malek, Roy Sooknanan, and Jean Compton

1. Introduction

Originally envisaged as an improved diagnostic method for detection of RNA viruses, NASBA™ has been developed into a technology with much wider application.

NASBA™ technology depends on selective primer-template recognition to drive a cyclic, exponential amplification of the target sequence *(1)*. Notably, NASBA™ operates continuously under isothermal conditions and consequently achieves rapid amplification using only standard laboratory equipment.

The fidelity and sensitivity of NASBA™ have both been demonstrated *(2)*. NASBA™ achieves amplification in the order of 10^9-fold in approx 90 min, with an error frequency of <0.3%. And in reconstructed clinical samples, e.g., fewer than 10 HIV molecules are detectable in a drop of human blood *(3)*. The use of nested primers in particularly complex samples augments the specific amplification of target sequences relative to the background.

One strength of the NASBA™ technology is in the direct amplification of RNA targets; a cDNA step is unnecessary and an RNA sequence can be selectively amplified in the presence of native genomic DNA. DNA samples are similarly amplifiable, however an additional priming

From: *Methods in Molecular Biology, Vol. 28:*
Protocols for Nucleic Acid Analysis by Nonradioactive Probes
Edited by: P. G. Isaac Copyright ©1994 Humana Press Inc., Totowa, NJ

step is necessary before the so-called "cyclic" phase of a standard NASBA™ reaction begins (*see* Fig. 1).

2. Materials

Reagent buffers should be made up in quantity and stored as aliquots at −20°C in a freezer that is not frost-free. Except where noted, reagents should be thawed on ice and remain on ice during use. Special care should be taken when preparing all reagents and reaction mixtures to eliminate the possibility of contamination (*see* Note 1). Company names in parentheses are recommended suppliers for each item. Alternative suppliers may be acceptable but have not been tested.

2.1. Amplification Materials

1. Primers: Synthetic primers should be gel purified, using a volatile buffer, such as triethylaminoacetate and methanol (in a 1:1 ratio), for extraction (*see* Note 2). A primer stock concentration of 100 pmol/µL is recommended. Primer 1 (P1) should be synthesized with the following general structure:

   5'-T7 Promoter sequence plus ~18 bases (complementary to target)-3'

   in which the recommended T7 RNA Polymerase promoter sequence is:

   5'-AATTCTAATACGACTCACTATAGGGAGA-3'

   Primer 2 (P2) should complement the cDNA sequence generated by reverse transcription of the target from P1. The target sequence to be amplified should be flanked by P1 and P2. The recommended length of P2 is approx 20 bases.
2. 2*M* Potassium chloride: Prepare solution and autoclave; readjust volume if necessary by adding sterile ultrapure water (*see* Note 3).
3. 1*M* Magnesium chloride (Sigma, St. Louis, MO): The use of commercially available, premade solution is highly recommended (*see* Note 4).
4. 1*M* Tris-HCl, pH 8.5 (Trizma-HCl, Sigma): The use of commercially available, premade solutions is highly recommended.
5. 25 m*M* Deoxynucleoside triphosphates (dNTPs; Pharmacia, Uppsala, Sweden): Combination of equal volumes of commercially available 100 m*M* stock solutions of dATP, dCTP, dGTP, dTTP.
6. 25 m*M* Nucleoside triphosphates (NTPs; Pharmacia): As above, using 100 m*M* stocks solutions of ATP, CTP, GTP, and UTP; also available commercially.
7. 250 m*M* Dithiothreitol (DTT, BDH associate of E. Merck, Darmstadt, Germany): Dissolve in sterile ultrapure water to make a 1*M* solution

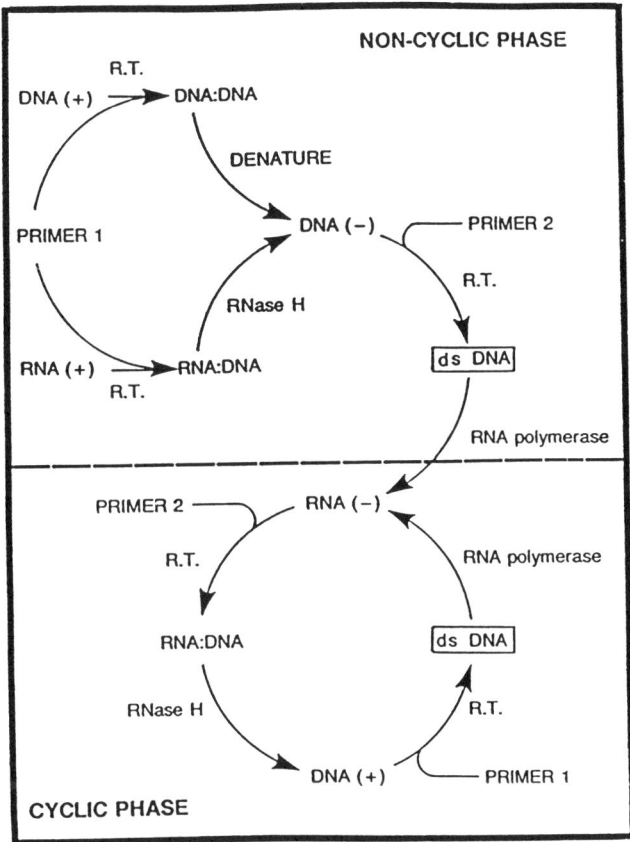

Fig. 1. Process for the continuous homogeneous isothermal amplification of RNA or DNA. To visualize the choreography of NASBA™ reactions, it is helpful to follow a single RNA molecule through the process. Noncyclic phase: (a) Primer 1 (P1) anneals to complementary sequence in RNA target; (b) Reverse Transcriptase (RT) extends the 3' end of P1, forming an RNA:DNA hybrid; (c) RNase H hydrolyzes the RNA strand of the hybrid, leaving single-stranded DNA (ssDNA), antisense to the original template; (d) P2 anneals to its complementary sequence within the new DNA strand; (e) RT extends the 3' end of the P2 primer, rendering the promoter region double-stranded; (f) T7 RNA polymerase generates RNA transcripts. Again, these RNA molecules are antisense to the original, and form templates for the cyclic phase of NASBA™. Cyclic phase: (g) Similar to steps a–f above, but characterized by primers binding in the reverse order. Reprinted, with permission, from ref. *1*.

and store as aliquots at –20°C. For use, dilute to 250 m*M* with sterile, ultrapure water; 250 m*M* solutions should not be stored longer than 6 mo.

8. 100% Dimethylsulfoxide (DMSO, BDH): Use as supplied (once thawed do not put on ice; DMSO will freeze). A freshly opened bottle should be used; aliquot and store at –20°C.

9. Bovine serum albumin (BSA; Boehringer Mannheim, Mannheim, Germany; Molecular Biology Grade): Available commercially as 20 mg/mL stock solution.

10. T7 RNA Polymerase (Pharmacia, *see* Note 5): Use as supplied; store at –20°C.

11. AMV reverse transcriptase (RT; Seikagaku America, Rockville, MD, *see* Note 5): Use as supplied; store at –20°C. Alternatively, *Bst* polymerase (Bio-Rad, Richmond, CA) can be used for DNA priming.

12. *E. coli* RNase H (Pharmacia, *see* Note 5): Use as supplied; store at –20°C.

13. Enzyme mixture (for single reaction, *see* Note 6): 20 U T7 RNA Polymerase (item 10 above), 8 U RT (item 11 above), 0.2 U RNase H (item 12 above), 2.5 µg BSA (item 9 above).

14. RNase Inhibitor (RNAguard™; Pharmacia): Use as supplied; store at –20°C.

15. Ultrapure water: Use sterile, nuclease-free water commercially available.

16. Concentrated (2.5X) NASBA™ buffer (to be supplied in kit or made from the above stocks): 100 m*M* Tris-HCl, pH 8.5, 125 m*M* KCl (*see* Note 3), 30 m*M* MgCl$_2$ (*see* Note 4), 5 m*M* each NTP, 2.5 m*M* each dNTP.

17. Primer mixture: This is an example of a suitable primer mixture for a single reaction (This can be scaled-up as necessary and stored at –20°C for at least 6 mo): 5 pmol Primer 1 stock, 5 pmol Primer 2 stock, 3.75 µL DMSO (100%) (*see* Note 7), Sterile Ultrapure water to give a final volume of 6.0 µL.

2.2. Detection Materials

1. TAE buffer, pH 8.0: 40 m*M* Tris-acetate; 1 m*M* EDTA, in deionized water. Adjust pH with acetic acid.

2. Ethidium bromide (Sigma): 10 mg/mL in TAE buffer.

3. Agarose gel: 3% (w/v) Nusieve GTG Agarose (FMC Bioproducts, Rockland, ME), 1% (w/v) agarose (Life Technolgies, Inc., Gaithersburg, MD), 0.5 µg/mL ethidium bromide in TAE buffer. Melt the agarose in TAE buffer, add water as necessary to adjust volume for evaporation taking place during melting, and cool before pouring. Add ethidium bromide stock solution when gel is almost cool enough to pour. Pour mixture onto prepared gel tray (with open ends sealed with a waterproof tape), add sample comb(s), and allow to cool thoroughly before use.

4. Running dye mixture: 0.05% (w/v) bromophenol blue (Life Technologies, Inc.) in 30% (v/v) glycerol (BDH), 10 mM EDTA (BDH, pH 8.0).

3. Methods

3.1. Standard NASBA™ Reaction

If the template is RNA, then perform the steps described here. If the template is DNA then perform the steps described in Section 3.2. first.

1. In the "clean" area (*see* Note 1), for each standard reaction add the following to a sterile 1.5-mL microcentrifuge tube: 8-A µL sterile ultrapure H$_2$O ($A = x + y + z$), 10.0 µL concentrated NASBA™ buffer (warmed; *see* Note 8), 1.0 µL 250 mM DTT, x µL (12.5 U) RNAguard™, y µL Enzyme mixture (*see* Section 2.), 6.0 µL Primer mixture. This makes a subtotal volume 25 – z µL.
2. Close each tube.
3. When all tubes are prepared, move to the template-handling area (*see* Note 9).
4. Add z µL template (RNA analyte or P1 primed DNA) to bring the total volume to 25 µL. Template volume will vary depending on concentration and desired number of input molecules, usually 10^4 molecules (*see* Note 10). For negative control reactions template volume will equal 0. (Summary of standard reaction conditions: 40 mM Tris-HCl, pH 8.5, 50 mM KCl, 12 mM MgCl$_2$, 2 mM NTPs (each), 1 mM dNTPs (each), 10 mM DTT, 0.2 µM P1, 0.2 µM P2, 15% DMSO, 100 µg/mL BSA, 20 U T7 RNA Polymerase, 8 U AMV Reverse Transcriptase, 0.2 U RNase H, 12.5 U RNAguard™, ~10^4 molecules of template).
5. Close each tube, gently mix by hand, and spin briefly in microcentrifuge to ensure all contents are in the bottom of the tube.
6. Incubate at 40°C for 60–90 min (*see* Note 11).
7. Analyze as desired, normally by agarose gel electrophoresis (Section 3.3., *see* Note 1).
8. The completed reaction mixture can be stored at –20°C for at least 3 mo.

3.2. Priming of DNA for Amplification

Follow these steps if a DNA template is to be used.

1. In the "clean" area (*see* Note 1) set up sterile microcentrifuge tubes for each reaction. Add the following components in this order: 9-x µL sterile ultrapure water, 8 µL concentrated NASBA™ buffer (warmed, *see* Note 8), 1 µL P1 primer (5 µM).
2. Close each tube and mix gently by hand.
3. Transfer to template-handling area.

4. To each tube add x µL deproteinized DNA sample, about 10^4 molecules (*see* Note 9).
5. Denature at 95°C for 5 min; incubate at 50°C for 5 additional min (for single-stranded DNA this step is not necessary).
6. Again in the template-handling area, add: 1 µL 100 mM DTT, 1 µL AMV Reverse Transcriptase (approx 10 U) (alternatively use 1 U of *Bst* DNA polymerase in place of RT).
7. Continue incubation at 50°C for 15 min (*see* Note 12).
8. Denature at 95°C for 5 min; chill immediately on ice. Keep on ice until use, or store at –20°C.
9. Use 2 µL of above prepared solution as a template in a standard NASBA™ reaction (*see* Section 3.1.).

3.3. Detection of NASBA™ Products

The products of the reaction are normally detected by agarose gel electrophoresis (*see* Section 2.2.). Reaction mixtures can also be analyzed by Northern transfer and hybridization using an oligonucleotide probe (e.g., *see* Chapters 7, 19, and 22).

1. Following the electrophoresis tank manufacturer's recommendations, pour the gel (*see* Section 2.2.). Prepare the 1X TAE electrophoresis buffer and add to the tank. Remove the sealing tape from the ends of the gel tray, and submerge the cooled gel in the buffer. Gently remove the comb(s).
2. Add 2 µL of running dye mixture to each completed NASBA™ reaction.
3. Load the samples into the wells carefully, including appropriate mol-wt markers.
4. Run the gel at 10 W (constant) for approx 1 h, or until the desired resolution is obtained.
5. Nucleic acid bands can be visualized on a UV transilluminator and photographed if desired (*see* Note 13).

4. Notes

1. As a minimum, the following guidelines are recommended in order to eliminate the possibility of contamination. NASBA™ reaction products must be kept physically separated from all NASBA™ reagents, including template; the use of a laminar flow hood equipped with UV lights is effective as a "clean" area for this purpose. Completed reactions should be opened with care in an area completely separated from other steps of the process. Neither template nor completed reactions should be handled in the "clean" area. A separate area for preparation and addition should be established; this area should have dedicated pipets and pipet tips.

Controls with no template should be set up entirely in the "clean" area, and not reopened until ready for analysis.

Sterile microcentrifuge tubes and pipet tips should be used for all reactions. These items should be sterilized by gamma irradiation rather than steam autoclaving. Sterile tips and tubes are available commercially (e.g., Bio-Rad [tips] and Sarstedt, Germany [tubes]). Sterile, cotton-plugged pipet tips are recommended to prevent carryover between samples and nuclease contamination from the pipetor. Pipet tips should not be used more than once.

Disposable gloves should be worn at all times while handling reagents and reaction products, and care should be taken not to crosscontaminate gloves during template addition. Fresh gloves should be used when preparing clean reagents or setting up reactions. Care should be taken to avoid nucleic acid and enzymatic contamination wherever possible (i.e., autoclave and filter [0.22 μm] all solutions, use disposable plasticware for storage and handling, and so on). Freedom from nuclease contamination is important even in the buffers used for analysis of products.

2. The use of gel-purified primers is recommended for all NASBA™ reactions. Primers should not be purified by ethanol precipitation, as this may increase the salt concentration in the reactions. Use a volatile buffer, such as triethylaminoacetate and methanol (1:1).

3. Monovalent cation concentration is critical; elevated levels in template preparations may inhibit the amplification reaction.

4. Mg^{2+} concentration is critical and must be kept equimolar with the combined concentration of deoxy- and ribonucleoside triphosphates. The use of quality-tested Mg^{2+} solutions is highly recommended.

5. Enzymes from different suppliers or different lots may vary; some adjustment of quantities may be necessary if mixtures are not already tested for NASBA™ use.

6. When making reagent mixtures, excess should be made to allow for pipeting losses. As a rule of thumb, at least 10% extra is recommended.

7. The DMSO concentration should not exceed 16–17% in the final mixture as it may inhibit the reaction. If the DMSO is too low (<15%) then little, or nonspecific, amplification may occur.

8. Concentrated NASBA™ buffer should be warmed until no longer cloudy before use.

9. Nucleic acid templates should be added to reactions last (in the designated template area). The recommended order of steps in standard reaction is: Assembly–Template Addition–Incubation–Analysis. All tubes should remain sealed once all reagents have been added, and be reopened for analysis only (in an area separate from all reagents and reaction assembly).

10. A total nucleic acid concentration of <1 µg is recommended.
11. Incubation temperatures should be kept constant at 40±0.5°C and monitored with a precision laboratory thermometer. The temperature should not be allowed to reach 41.5°C at any time.
12. If *Bst* DNA polymerase is used in place of RT, then the priming temperature should be 60°C.
13. If nonspecific products, with little or no specific amplification, occur:
    a. Primer–primer homology may have lead to dimerization; computer programs are available that predict possible homologies, allowing selection of alternative primer sequences.
    b. Template is either absent or is present in low amounts (≤10 copies), favoring primer–primer interactions.
    c. Primer(s) have misprimed at alternative sites; again, selection of new primer sequences is recommended.
    d. In DNA samples: Prepriming reaction may not have been sufficiently stringent. The priming temperature can be increased to 60°C if using *Bst* DNA polymerase instead of RT.
    e. Low- and high-mol-wt, nonspecific products (often smeary after gel electrophoresis) may be caused by lower than optimum monovalent cationic concentration. Adjust carefully with the addition of KCl if necessary.
    f. DMSO concentration may have been <15%; use fresh or stably stored DMSO solution.

References

1. Compton, J. (1991) Nucleic acid sequence-based amplification. *Nature* **350,** 91,92.
2. Malek, L., Darasch, S., Davey, C., Henderson, G., Howes, M., Lens, P., and Sooknanan, R. (1992) Application of NASBA™ isothermal nucleic acid amplification method to the diagnosis of HIV-1. *Clin. Chem.* **38,** 458.
3. Kievits, T., van Gemen, B., van Strijp, D., Schukkink, R., Dircks, M., Adriaanse, H., Malek, L., Sooknanan, R., and Lens, P. (1991) NASBA™ isothermal nucleic acid amplification optimized for the diagnosis of HIV-1 infection. *J. Virolog. Methods* **35,** 273–286.

Index